Following the Proper Channels

International Water Law Series

Series Editor

Stephen C. McCaffrey

VOLUME 6

The titles published in this series are listed at *brill.com/iwl*

Following the Proper Channels

Tributaries in the Mekong Legal Regime

By

Bennett L. Bearden

BRILL
NIJHOFF

LEIDEN | BOSTON

Library of Congress Cataloging-in-Publication Data

Names: Bearden, Bennett L., author.
Title: Following the proper channels : tributaries in the Mekong legal regime / By Bennett L. Bearden.
Description: Leiden : Brill, 2018. | Series: International water law series ; 6 | Includes bibliographical references and index.
Identifiers: LCCN 2017061430 (print) | LCCN 2018000752 (ebook) | ISBN 9789004362604 (eBook) | ISBN 9789004362598 (hardback : alk. paper)
Subjects: LCSH: Water resources development--Law and legislation--Mekong River Watershed. | Water-supply--Mekong River Watershed. | Mekong River Watershed--International status.
Classification: LCC KNC660.M45 (ebook) | LCC KNC660.M45 B43 2018 (print) | DDC 346.5904/6912--dc23
LC record available at https://lccn.loc.gov/2017061430

Typeface for the Latin, Greek, and Cyrillic scripts: "Brill". See and download: brill.com/brill-typeface.

ISSN 2351-9606
ISBN 978-90-04-36259-8 (hardback)
ISBN 978-90-04-36260-4 (e-book)

This book is printed on acid-free paper and produced in a sustainable manner.

To my wife, Amy, the only compass I need,
and my daughters, Brit and Morgan,
and
To Professor Steve McCaffrey, J.D., Dr. jur., the one who
taught me all things about water law and policy.

Contents

Acknowledgments

This book marks the culmination of a decade (2007–2017) of research in the area of transboundary water law and policy in the Mekong River basin. I spent four of those years (2007–2011) focusing on the legal and policy issues posed by the marginalization of tributaries in the Mekong legal regime during dissertation research for my Doctor of Juridical Science (JSD) in International Water Resources Law degree (legal education's counterpart to the PhD degree) at McGeorge School of Law, University of the Pacific, in Sacramento, California. I am indebted to my dissertation advisor, distinguished professor and scholar Stephen C. McCaffrey, Director of the International Water Resources Law Program, for guiding my academic journey and allowing me to study under his direction. Because of his willingness to mentor me and largely due to the prestigious water law program at McGeorge School of Law, I learned to think like a water lawyer and water policy analyst. Professor McCaffrey, one of the foremost authorities in the world on international water law, was unselfish with his time despite an often hectic schedule. His sage advice, insightful guidance, and unwavering support ensured the completion of this book.

My deepest gratitude is extended to my distinguished JSD Dissertation Committee—Committee Chair and External Examiner, Flavia Loures, who was formerly the Senior Program Officer and International Water Lawyer with the World Wildlife Fund (WWF) Global Freshwater Program and Global Faculty Member at the Centre for Water Law, Policy and Science at the University of Dundee, Scotland, UK; Professor Greg Weber, former Director of the Water Resources Law Program at McGeorge School of Law; and Professor Rachael Salcido, Director of the Environmental Law Program and Sustainable Development Institute at McGeorge School of Law. Their comments and suggestions about my research were invaluable.

The patience, support and kindness of former Dean Elizabeth Rindskopf Parker, Carol Hunter, former Registrar, and Rachel Dubay, former Coordinator of Graduate and International Programs, all of McGeorge School of Law, will not soon be forgotten. They created a law school environment where students are not only challenged, but also cared for. These gracious "caretakers" opened their offices and resources to me whenever I needed assistance during my doctoral residency and research. I would have been lost without them.

In addition, I would like to thank my colleagues in the Graduate International Water Group (GIWG), International Water Resources Law Program, at McGeorge School of Law, Vera Bezdicek, Ram Dhakal, Dan Fitzgerald, Ekin Kaya, Maria Milanes-Murcia, Rodrigo Rondon, Marjorie Simington, Yuming

Sun, Margaret Vick and Wen-Lu Wang, for their friendship and insightful suggestions and comments about my research.

Special thanks are extended to Dr. Aaron Wolf, Professor and Director of the Program in Water Conflict Management and Transformation and the Transboundary Freshwater Dispute Database (TFDD), Oregon State University, Department of Geosciences, and Dr. Avijit Gupta, School of Geography, University of Leeds, UK, for their assistance during my research.

Special thanks are extended to Irene Burgess, Executive Secretary at the Geological Survey of Alabama, and Nathaniel Broadhurst, Research Assistant and Policy Analyst at The University of Alabama Water Policy and Law Institute, for their assistance in the research and preparation of this book. They made invaluable contributions to this work.

I owe a large debt of gratitude to my loving parents, Benjamin E. Bearden (1931–2008) and Londa Bennett Bearden, who showed me the way. Finally, special thanks to my beloved wife, Amy Reele Horton Bearden, whose love, support, patience, friendship, and encouragement made my JSD research possible.

Abstract

In 1957, the four lower Mekong River basin (LMRB) states jointly organized the development of the lower basin and established a legal regime that has spanned six decades of more or less successful cooperation. In 1995, the LMRB countries concluded the Agreement on the Cooperation for the Sustainable Development of the Mekong River basin (1995 Mekong Agreement).[1] Lauded at the time as the most progressive of institutional frameworks for the governance of an international watercourse, the 1995 Mekong Agreement has, at its core, the concepts of cooperation and sustainable development[2] and was the first transboundary watercourse agreement to adopt the latter paradigm.[3] This multilateral treaty, which emphasizes joint basin-wide development and environmental protection, established the robust and well-funded Mekong River Commission (MRC), the basin's joint institution. Seemingly, the so-called "Mekong spirit"[4] of cooperation is resilient to any political turmoil in the lower basin and is heralded as a model for the rest of world.[5] Guided by the 1995 Mekong Agreement's sustainable development paradigm, the MRC attempts to harmonize and balance the maintenance of water quantity with protection of water quality. The parties agree to cooperate and use the Mekong's water resources in a manner in which the river system's environmental conditions and ecological balance are conserved and maintained.

Despite its lofty objectives, the treaty has significant limitations, not least in its failure to legally require "holistic water resources management"[6] and

1 Chiang Rai, Thailand, 5 April 1995, 34 ilm 864 (1995).

2 *See* generally, Jeffrey W. Jacobs, *Towards sustainability in lower Mekong River basin development*, 19 Water International 43 (1994); Mekong River Commission [hereinafter MRC, Mekong River Commission: towards sustainable development, Annual Report (1995); Hiroshi Hori, The Mekong: Environment and Development, United Nations University Press, Tokyo (2000) for discussions of the sustainable development paradigm in the Mekong River Basin.

3 Bantita Pichyakorn, 'Sustainable Development and International Watercourse Agreements: The Mekong and the Rhine' (International Union for Conservation of Nature (IUCN) Draft Paper, June 30, 2002) 14.

4 U. Nyun described the "Mekong spirit" as "the great goodwill, the friendly spirit of collaboration, the abundant enthusiasm which animates Mekong work." See generally, P.K. Menon, "Financing the Lower Mekong River Basin Development" (1972) 44 Pacific Affairs 566.

5 MRC, *The Story of Mekong Cooperation* <http://www.mrcmekong.org/about-the-mrc/history/> (accessed 12 June 2017).

6 Stephen C. McCaffrey & Gregory S. Weber, *Guidebook for Policy and Legislative Development on Conservation and Sustainable Use of Freshwater Resources* (United Nations Environment Programme, 2005).

promote *basin-wide* "sustainable water development."[7] Perhaps its most glaring shortcoming is the treaty's marginalization of tributaries and truncation of headwaters. This renders the Mekong the archetypal *dissevered river basin* that on paper, and legally, is in effect separated, divided into parts, and broken up. In a blizzard of undefined and ambiguous terms used throughout the treaty, its text generalizes and redefines the concept of a river basin;[8] yet on paper this vaunted Agreement creates the legal fiction of a holistic water resources management paradigm by clinging fervently to the "imported"[9] concept of sustainable development. In effect, tributaries in the Mekong legal regime are managed by each state unilaterally or at best, bilaterally, through application of the limiting language embodied in Article 5(A), the tributary rule, which emphasizes hydrosovereignty.

The modern trend and badge of responsibility under the 1997 UN Convention on the Law of the Non-Navigational Uses of International Watercourses (UN Watercourses Convention), which entered into force in 2014, are to provide a legal framework for the entire watercourse, defined to include its tributaries,[10] and to promote basin-wide governance. The Xayaburi Dam planned by the Lao PDR on the mainstem of the Mekong illustrates the Achilles heel of the Mekong treaty, which evidently lacks effective mechanisms to prevent even the most brazen unilateral development of the River. In the Mekong legal regime, the limited extent to which planned measures and megaprojects on tributaries and even the mainstream are subject to interdependence and basin-wide governance remains problematic. More than 20 years of experience has shown that there are aspects of the 1995 Mekong Agreement that should be strengthened in order to secure the environmental, economic, and social benefits it promises. A review of the 1995 Mekong Agreement and the UN Watercourses Convention *in pari materia* reveals that these two treaties are generally compatible. The entry into force of the UN Watercourses Convention in August 2014

7 Pichyakorn, *supra* note 3, at 10.

8 A river basin is defined as "the entire area drained by a river *and its tributaries*." R.L. Bates and J.A. Jackson (eds.), *Glossary of Geology* (American Geological Institute, 1980). (emphasis added).

9 Bastien Affeltranger, Intra-basin conflict resolution in the Mekong basin: Is a reconciliation of water values possible? 114 *in* International workshop on the value of water: Different approaches in transboundary water management, Koblenz (10–11 March 2005).

10 Stephen C. McCaffrey, *The Law of International Watercourses*, (Oxford University Press, 2nd ed., 2007) 35.

opens new vistas for managing Mekong tributaries. Indeed, Vietnam was the 35th state to become a party to the UN Watercourses Convention, triggering the treaty's entry into force. Cambodia, the Lao People's Democratic Republic, and Thailand should now join the UN Watercourses Convention thereby strengthening the relationship between the 1995 Mekong Agreement and customary international law and demarginalizing Mekong tributaries.

List of Illustrations

Figures

Tables

List of Abbreviations

BDS	Basin Development Strategy
billion m^3	Billion cubic meters
CEO	MRC Secretariat Chief Executive Officer
ECAFE	Economic Commission for Asia and the Far East
EIA	Environmental Impact Assessment
ESCAP	Economic and Social Commission for Asia and the Pacific
EVN	Electricity of Vietnam
GA	General Assembly of the United Nations
GMS	Greater Mekong Subregion
ha	Hectare(s)
ICJ	International Court of Justice
ILA	International Law Association
ILC	International Law Commission
IRN	International Rivers Network
IUCN	International Union for Conservation of Nature
IWRM	Integrated water resources management
JC	Joint Committee of the MRC
km	Kilometer(s)
km^2	Square kilometers
km^3/yr	Cubic kilometer per year
LMI	Lower Mekong Initiative
LMCM	Lancang-Mekong Cooperation Mechanism
LMRB	Lower Mekong River Basin
m	Meter(s)
$m^3/capita/yr$	Cubic meters per capita per year
m^3/day	Cubic meters per day
m^3/sec (m^3/s^{-1})	Cubic meters per second
$m^3/year$	Cubic meters per year
million m^3	Million cubic meters
mm	Millimeters
MRC	Mekong River Commission
MW	Megawatts
MWG	Mekong Working Group
NGO	Non-governmental Organization
NMC	National Mekong Committee
PCIJ	Permanent Court of International Justice
PDR	People's Democratic Republic

PNPCA	Procedures for Notification, Prior Consultation, and Agreement
RID	Thai Royal Irrigation Department
SPN	Sesan Protection Network
SWECO	"Swedish Consultants"
TDM	Technical Drafting Meeting
3SPN	3S Rivers Protection Network
3S RIVERS	Sesan, Sekong and Srepok rivers
UK	United Kingdom of Great Britain and Northern Ireland
UMRB	Upper Mekong River Basin
UN	United Nations
UNDP	United Nations Development Programme
UNEP	United Nations Environment Programme
UNCLOS	United Nations Convention on the Law of the Sea
UNESCO	United Nations Educational, Scientific and Cultural Organization
UN GAOR	United Nations General Assembly Official Records
US/U.S./USA	United States of America
USAID	United States Agency for International Development
U.S.C.	United States Code
WWF	World Wildlife Fund
YBILC	Yearbook of the International Law Commission

List of Legislative Materials

International Instruments

Treaty between the United States and Great Britain Relating to Boundary Waters and Questions Arising between the United States and Canada, opened for signature 11 January 1909, 23 UKTS 1910 (entered into force 5 May 1910) (1909 Boundary Waters Treaty).

Treaty of Peace between the Allied and Associated Powers and Germany, Versailles, 28 June 1919, 11 Martens, 3rd ser., p. 323 (Treaty of Versailles).

Statute for the Committee for Co-Ordination of Investigations of the Lower Mekong Basin (Cambodia, Laos, Republic of South Viet-Nam and Thailand), Bangkok, Thailand, 17 September 1957 (1957 Mekong Statute).

Agreement between Bolivia and Peru concerning a Preliminary Economic Study of the Joint Utilization of Lake Titicaca of September 19, 1957, in Legislative Texts, Treaty no. 45, p. 168, ratified by Peru in 1957 and by Bolivia in 1987.

Treaty Relating to Cooperative Development *of the Water Resources of the Columbia River Basin* signed at Washington, 17 January 1961, and exchanges of notes at Washington, 22 January 1964, and at Ottawa, 16 September 1964, 15 U.S.T. 1555, 542 U.N.T.S. 244 (1961 Columbia River Treaty).

Proposed (unratified) 1966 Amendments to 1957 Mekong Committee Statute, UN ECAFE, Mekong Committee, 29th Session (Special), 12–13 August 1965, revised 29 April 1966, Vientiane, Laos.

Helsinki Rules on the Uses of the Waters of International Rivers, adopted by the International Law Association at the 52nd Conference held in Helsinki, Finland, 14–20 August 1966, Report of the 52nd Conference, 484–532 (1966) (1966 Helsinki Rules).

Vienna Convention on the Law of Treaties, opened for signature 23 May 1969, entered into force 27 January 1980, 1155 U.N.T.S. 331, 8 ILM 679 (1969) (1969 Vienna Convention).

Draft Lower Mekong Basin Water Charter (with Commentary), Mekong Committee (Khmer Republic, Laos, Thailand and the Republic of South Viet-Nam), UN Economic Commission for Asia and the Far East (ECAFE), Working Paper by the Secretariat, Doc. WRD/MKG/INF/L.427, Bangkok, Thailand, 20 March 1971 (Water Charter).

Draft Statement of Principles for the Planning and Utilization of the Waters of the Lower Mekong Basin, Mekong Committee, WRD/MKG/INF/L.466, Bangkok, Thailand, 1971 (Draft Statement).

Joint Declaration of Principles for Utilization of the Waters of the Lower Mekong Basin, Mekong Committee (Khmer Republic, Laos, Thailand and the Republic of Vietnam), Vientiane, Laos, 31 January 1975 (1975 Joint Declaration).

Declaration Concerning the Interim Committee for Coordination of the Lower Mekong Basin, Vientiane, Laos, 5 January 1978 (1978 IMC Declaration).

Convention Creating the Niger Basin Authority, signed at Faranah on 21 Nov. 1980, available at TFDD at <http://ocid.nacse.org/tfdd/tfdddocs/418ENG.pdf> (accessed 29 July 2010).

UN Convention on the Law of the Sea, 10 December 1982, UN Doc. A/CONF.62/122, 21 ILM 1261 (1982) (UNCLOS).

Agreement Between the Federal Republic of Nigeria and the Republic of Niger Concerning the Equitable Sharing in the Development, Conservation and Use of Their Common Water Resources, 18 July 1990 (1990 Nigeria-Niger Tributary Agreement) available at the Transboundary Freshwater Dispute Database (TFDD) at <http://ocid.nacse.org/tfdd/tfdddocs/483ENG.pdf> (accessed 29 July 2010).

1991 European Convention on Environmental Impact Assessment in the Transboundary Context, 25 February 1991, 1989 U.N.T.S. 310, entered into force 10 September 1997 (Espoo Convention).

Agreement of September 14, 1992, between Namibia and South Africa on the Establishment of a Permanent Water Commission, 32 ILM 1147 (1993).

Treaty Concerning the Construction and Operation of the Gabcikovo-Nagymaros System of Locks, 32 ILM 1247 (1993) (Gabcikovo-Nagymaros Treaty).

Agreement on the Cooperation for the Sustainable Development of the Mekong River Basin, Chiang Rai, Thailand, 5 April 1995, 34 ILM 864 (1995) (1995 Mekong Agreement).

Protocol on Shared Watercourse Systems in the Southern African Development Community Region, 28 August 1995, entered into force 29 Dec. 1998 *in Treaties Concerning the Non-Navigational Uses of International Watercourses, Africa,* FAO Legislative Study 61, p. 146, at pp. 147–148 (1997) (SADC Protocol).

1997 UN Convention on the Law of the Non-Navigational Uses of International Watercourses, 21 May 1997, U.N. Doc. A/RES/51/869, *reprinted in* 36 ILM 700 (1997) (1997 UN Convention).

1998 Convention on Access to Information, Public Participation in Decision Making and Access to Justice in Environmental Matters, 25 June *1998,* Aarhus, Denmark, entered into force 30 October 2001, available at <http://www.unece.org/env/pp/documents/cep43e.pdf> (accessed 29 July 2010) (Aarhus Convention).

Agreement on Commercial Navigation along the Upper Reach of the Mekong River, (China, Lao PDR, Myanmar and Thailand) (April 2000).

2002 Water Charter of the Senegal River, Charte des Eaux du Flevre Senegal, adopted by the Conference of Heads of State and Government, Organisation pour la Mise en Valeur Du Fleuve Senegal (OMVS), 18 May 2002, OMVS Resolution 005, available at <http://lafrique.free.fr/traile/omvs_200205.pdf> (text in French, unofficial English translation on file with author) (accessed 14 March 2008).

Berlin Rules on Water Resources, International Law Association, Report of the 71st Conference 3 (2004), 71 ILA 337, 385 (2004) (Berlin Rules).

International Law Commission, *Draft Articles on the Law of Transboundary Aquifers*, 2008, UN GAOR, 63rd session, Supp. No. 10, UN Doc. A/63/10, available at <http://untreaty.un.org/ilc/texts/instruments/english/draft%20articles/8_5_2008.pdf>.

Cambodia

Law on Water Resources Management, National Assembly, 29 June 2007.

Lao PDR

Policy on Water and Water Resources.

Law on Water and Water Resources, National Assembly, 11 October 1996.

Thailand

Water Resources Bill (draft Water Act).

United States

1922 Colorado River Compact, signed at Santa Fe, New Mexico, 24 November 1922, H. Doc. 605, 67th Cong., 4th Sess., pp. 8–12 (1923), 70 Cong. Rec. 324 (1968).

1922 La Plata River Compact, 27 Nov. 1922, available at <http://www.ose.state.nm.us/isc_laPlata_river_compact.html> (accessed 2 July 2010).

1928 Boulder Canyon Project Act, 43 U.S.C. sec. 617 *et seq.* (1928 BCPA).

1948 Upper Colorado River Basin Compact, 63 Stat. 31 (1949).

Clean Water Act, 33 U.S.C. sec. 1251 *et seq.* (1972).

U.S. Council on Environmental Quality Regulations, 40 CFR 1508.7 and 1508.8.

ARIZ. REV. STAT. sec. 645–1311 (2006) (State of Arizona).

Vietnam

Law on Water Resources, National Assembly, May 1998.

Government Decree on Functions, Tasks, Authorities and Organizational Structure of Ministry of Agriculture and Rural Development, Vietnam (2003).

List of Cases

International Courts

Territorial Jurisdiction of the International Commission of the River Oder, Judgment No. 16, 10 Sept. 1929, PCIJ Ser. A, No. 23, pp. 5–46. Reproduced in Manley O. Hudson, ed., *World Court Reports*, vol. 2, p. 609 (1969) (*River Oder* case).

1996 Advisory Opinion on the Legality of the Threat or Use of Nuclear Weapons, ICJ 226 (8 July).

Case concerning the Gabcikovo-Nagymaros Project (Hungary/Slovakia) Judgment of September 25, 1997, 1997 ICJ 7, *reprinted in* 37 ILM 162 (1998), also available on the ICJ website, <www.icj-cij.org> (*Gabcikovo-Nagymaros* case).

Case Concerning Pulp Mills on the River Uruguay (Argentina v. Uruguay), Judgment of 20 April 2010, available on the ICJ website, <www.icj-cij.org> (*Pulp Mills* case).

Germany

Wurttemberg and Prussia v. Baden (*The Donauversinkung Case*), German Staatsgerichtshof, 1927, Entscheidungen des Reichesgereichts in Zivilsachen, v. 116, appendix, 18–45, 86 Ann. Dig. 128 at 131.

United States

Arizona v. California, 373 U.S. 546 (1963).
Driscoll v. Adams, 181 F.3d 1285 (11th Cir. 1999).
Irwin v. Phillips, 5 Cal.140 (1855).
Kansas v. Colorado, 206 U.S. 46 (1907).
New Jersey v. New York, 283 U.S. 336 (1931).

Arbitral Decisions

Trail Smelter case (US v. Canada) 3 UNRIAA 1911, 1938, 1963, 1966, 3 R. Int'l Arb. Awards 1905, 1965 (1941) (*Trail Smelter* arbitration).

Lake Lanoux Arbitration (France v. Spain), 12 UNRIAA 281, 62 RGDIP 79 (1958), both in the original French. For English translations, see 24 ILR 101 (1961), 53 AJIL 156 (1959), and YILC, vol. 2, Part 2, p. 194 (1976) (*Lake Lanoux* arbitration).

Introduction

In 1957, the four lower Mekong River basin (LMRB) states—Cambodia, the Lao People's Democratic Republic (Lao PDR), Thailand and Vietnam—jointly organized the development of the basin and established a legal regime that has spanned six decades of cooperation. In 1995, the LMRB countries concluded the Agreement on the Cooperation for the Sustainable Development of the Mekong River Basin (1995 Mekong Agreement).[1] Lauded as the most progressive of institutional frameworks for the governance of an international watercourse, the 1995 Mekong Agreement has the concept of sustainable development[2] at its core and was the first transboundary watercourse agreement to adopt this paradigm.[3] This multilateral treaty, which emphasizes joint basin-wide development and environmental protection, established the robust and well-funded Mekong River Commission (MRC), the basin's joint institution. Seemingly, the so-called "Mekong spirit"[4] of cooperation is resilient to any political turmoil in the basin and is heralded as a "model for the world."[5] Guided by the 1995 Mekong Agreement's sustainable development paradigm, the MRC attempts to harmonize and balance the maintenance of water quantity with protection of water quality. The LMRB states agree to cooperate and use the Mekong's water resources in a manner in which the river system's environmental conditions and ecological balance are conserved and maintained.

Despite its lofty objectives, the treaty has significant limitations, not least in its failure to legally require "holistic water resources management"[6] and

---

1 Chiang Rai, Thailand, 5 April 1995, 34 ILM 864 (1995) [hereinafter 1995 Mekong Agreement].

2 For discussions of the sustainable development paradigm in the Mekong River Basin, *see generally,* Jeffrey W. Jacobs, "Towards Sustainability in Lower Mekong River Basin Development" (1994) 19 *Water International* 43; Mekong River Commission [hereinafter MRC], "Mekong River Commission: Towards Sustainable Development" (Annual Report, 1995); Hiroshi Hori, *The Mekong: Environment and Development* (United Nations University Press, 2000).

3 Bantita Pichyakorn, "Sustainable Development and International Watercourse Agreements: The Mekong and the Rhine" (International Union for Conservation of Nature (IUCN) Draft Paper, 30 June 2002) 14.

4 U. Nyun described the "Mekong spirit" as "the great goodwill, the friendly spirit of collaboration, the abundant enthusiasm which animates Mekong work." *See generally,* P.K. Menon, "Financing the Lower Mekong River Basin Development" (1972) 44 *Pacific Affairs* 566.

5 MRC, *The Story of Mekong Cooperation* <http://www.MRCmekong.org/about-MRC/history/> (accessed 12 June 2017).

6 Stephen C. McCaffrey and Gregory S. Weber, *Guidebook for Policy and Legislative Development on Conservation and Sustainable Use of Freshwater Resources* (United Nations Environment Programme [hereinafter UNEP], 2005).

 | DOI 10.1163/9789004362604_002

promote *basin-wide* "sustainable water development."[7] The absence of China and Myanmar (Burma) from the cooperative framework of the 1995 Mekong Agreement has important governance implications. There is no clear definition of sustainable development in the text, and environmental protection mechanisms fall short of specific international customary laws. The 1995 Mekong Agreement provides only a skeletal framework for equitable allocation and maintenance of *mainstream* flows leaving the details of the latter to mere procedures of questionable legality. Perhaps the most glaring deficiency in the Agreement's effectiveness derives from its passive regulation of tributaries, which allows countries to unilaterally divert and transfer water without approval from other basin states. This regime passivity is especially important now because "[b]y 2010, more than 120 tributary and mainstream projects [were] projected, under construction or operating on lower Mekong tributaries, in addition to the 11 proposed for the Mekong mainstream."[8] Development of the river's tributaries has rendered the efficacy of the Mekong legal regime to support holistic water resources management and basin-wide sustainable water development questionable. Indeed, the Xayaburi Dam planned by the Lao PDR on the mainstream of the Mekong illustrates the Achilles heel of the Mekong treaty, which evidently lacks effective mechanisms to prevent even the most brazen unilateral development of the river. Although the Agreement's objectives are laudable, policy implementation has been disappointingly slow if harmonization of water use and protection, sustainable water development, and basin-wide governance are indeed its goals.

Tributaries constitute part of a unitary whole of a river basin.[9] More than conveyances of water, tributaries are complex networks of hydrological, geological, biological, chemical, physical and geographical interconnectivity. Management of tributaries is a subject of increasing environmental

---

7 Pichyakorn, *supra* note 3, at 10.

8 Philip Hirsch, "The Changing Political Dynamics of Dam Building on the Mekong" (2010) 3(2) *Water Alternatives* 312, 317 (citing the MRC, *Hydropower database* (2010)).

9 Stephen C. McCaffrey, *The Law of International Watercourses*, (Oxford University Press, 2nd ed., 2007) 35 [hereinafter McCaffrey]. McCaffrey explains that the "[t]he expression "watercourse" is defined in the 1997 UN Convention… as "a system of surface waters and groundwaters constituting by virtue of their physical relationship a unitary whole and normally flowing into a common terminus.… This hydrologic system consists of a number of components, including rivers and their tributaries.… As implied by the word "system", these components are interrelated with one another." (internal citations omitted). This definition mirrors that adopted by the International Law Commission (hereinafter ILC). Draft articles on the *Law of the Non-Navigational Uses of International Watercourses*, Yearbook of the International Law Commission [hereinafter YBILC], vol. 2, pt. 2, at 90 (1997).

scrutiny in the LMRB. As is the case with some large rivers, half of the water flowing in the Mekong is derived from its tributaries, but "marginalization"[10] of these important riverine arteries in the Mekong legal regime is leading to unilateral and bilateral water diversions. These diversions alter flow regimes, fragment habitats, reduce sediment transport and channel maintenance, reduce water quality, and cause cumulative effects. Seasonal and annual variations in flow are essential to provide energy, habitat, and biodiversity in a flood pulse river such as the Mekong. Flooding connects Mekong tributaries to their floodplains, providing feeding grounds and escape from prey for aquatic biota while allowing exchange of organic materials between tributaries and wetlands. Impoundments and water diversions on Mekong tributaries cause habitat fragmentation altering aquatic ecosystem connectivity and hindering important natural functions such as fish migration and spawning. In their natural state, Mekong tributaries form channels that balance sediment supply and water runoff. Reservoirs on Mekong tributaries interrupt the downstream transport of sediment and floodwaters, resulting in a stream channel that is no longer in balance with its sediment and flow regime. Altered flow regimes decrease a tributary's ability to assimilate waste and maintain water quality. Cumulative impacts, defined as "the impact on the environment which results from the incremental impact of the action when added to other past, present, and reasonably foreseeable future actions regardless of what agency … or person undertakes such other action,"[11] have not yet begun to be fully studied, much less understood, on Mekong tributaries. Analysis of selected international and interstate legal regimes,[12] including the multilateral Mekong treaty, demonstrates that despite their significant volumetric contributions and other

10 Naomi Carrard, "Mainstream or Marginal? Transboundary Tributaries and the Mekong Agreement" (Presentation delivered at the 8th International Riversymposium, Brisbane, September 2005).

11 *See, e.g.*, 40 CFR 1508.7; The U.S. Council on Environmental Quality regulations (40 CFR §§ 1508.7 and 1508.8) define the impacts and effects that must be addressed and considered by Federal agencies in satisfying the requirements of the National Environmental Policy Act process, which includes direct, indirect, and cumulative impacts. Cumulative impacts can result from individually minor but collectively significant actions taking place over a period of time.

12 Stephen D. Krasner, "Structural Causes and Regime Consequences: Regimes as Intervening Variables" in Stephen D. Krasner (ed.), *International Regimes* (Cornell University Press, 1986) 2. Krasner defines a regime as "sets of implicit or explicit principles, norms, rules, and decision-making procedures around which actors' expectations converge in a given area of international relations." This definition is widely accepted and used in the legal and policy literature.

important natural functions, ironically, tributaries are often marginalized and treated differently from the mainstream in many of these regimes.[13]

In the past two decades, large-scale hydropower and water diversions on Mekong tributaries have been the focus of considerable political attention. These "megaprojects"[14] have moved up on most Mekong countries' national agendas, with an attendant increase in sectoral management of these planned measures. Emblematic of this sectoral management is the marginalization of tributaries in the Mekong legal regime. Already, unbridled hydropower projects and water diversions on Mekong tributaries are giving rise to interstate conflicts in the LMRB, exacerbating potential water quantity and quality problems, increasing uncertainty about water availability, and replicating the legal dynamics played out in other international basins. In this process, basin-wide governance envisioned in the 1995 Mekong Agreement has been supplanted by national sovereignty that reduces regional cooperation between states sharing tributaries. Hydrosovereignty, the quality of having supreme, independent authority over tributaries by construction of socio-politically constructed boundaries, cannot be reconciled with the physical boundaries imposed by hydrology. Managing hydrointerdependency is at the core of the

13 *See* Treaty between the United States and Great Britain Relating to Boundary Waters and Questions Arising between the United States and Canada, opened for signature 11 January 1909, 23 UKTS 1910 (entered into force 5 May 1910), preliminary art. [hereinafter referred to as 1909 Boundary Waters Treaty]. For a history of the 1909 Boundary Waters Treaty and international water relations between the US and Canada, see McCaffrey, *supra* note 9, at 346–356; Treaty relating to Cooperative Development of the Water Resources of the Columbia River Basin, signed at Washington, 17 January 1961, and exchanges of notes at Washington, 22 January 1964, and at Ottawa, 16 September 1964, 15 U.S.T. 1555, 542 U.N.T.S. 244 [hereinafter referred to as 1961 Columbia River Treaty]; Interstate water disputes often replicate the legal dynamics played out in international water conflicts. Similarly, interstate acts and judicial decisions may mirror international agreements in their disparate treatment of tributaries. In *Arizona v. California*, 373 U.S. 546 (1963), the central issue in the case concerned whether each state could divert water from tributaries of the Colorado River. California, seeking more water, targeted Arizona's Gila River, a Colorado tributary. Arizona claimed that the apportionment of the Colorado River Waters in the 1928 Boulder Canyon Project Act, 43 U.S.C. § 617 *et seq.* included only mainstream waters while California contended that the apportionment included all tributary waters. The U.S. Supreme Court ruled that the 1928 Boulder Canyon Project Act reserved to Arizona, California, and Nevada the exclusive use of the waters of each state's own tributaries.

14 *See generally,* Francois Molle and Philippe Floch, "Water, Poverty and the Governance of Mega Projects: The Thai 'Water Grid'" (IWMI Working Paper, 2007); Francois Molle and Philippe Floch, "Megaprojects and Social and Environmental Changes: The Case of the Thai 'Water Grid'" (2008a) 37 *Ambio* 199. for discussions of "megaprojects."

strife internalized in the 1995 Mekong Agreement. Indeed, internal political forces often lead the Mekong states to maximize their use of the Mekong River system without regard to the needs of their co-riparians, thereby vitiating the importance of the basin concept asserted by the states in the 1995 Mekong Agreement, as well as their obligations under generally accepted legal norms governing the use of water in shared tributaries. Sovereignty over transboundary tributaries is not absolute. Because of the rift between sovereignty and hydrology, as well as the importance of the large volumetric flow contributions of Mekong tributaries to the mainstream, tributary watercourse law deserves more attention.

Arguably, through redefinition of the term "basin," and the use of other undefined and ambiguous terms, tributaries are effectively taken out of the Mekong legal regime for management purposes or, in the best case scenario, marginalized. This study examines the tendency of the parties to the 1995 Mekong Agreement to circumvent holistic water resources management and sustainable water development in pursuit of national agendas through the unilateral and bilateral development of tributaries. The overarching legal and hydropolicy issue is whether states can simultaneously pursue hydrosovereignty on tributaries and ensure the Mekong legal regime's efficacy to achieve holistic water resources management and basin-wide governance. Key questions to consider are (1) whether states can simultaneously pursue national sovereignty agendas on tributaries in a river basin *system* and comply with international laws and regional policies that would best ensure a multilateral legal regime's survival; and (2) when tributaries are marginalized in that multilateral regime, whether states can cope with the challenges and conflicts of unilateral and bilateral development in the use of these river arteries, shared by countries which are at different stages of political and economic development and which have disparate interests. In addressing this issue and the attendant questions, marginalization of tributaries is analyzed in the context of the law of international watercourses law and policy, hydrosovereignty, and the national economic development interests of the concerned riparians.

Chapter 1 of this book describes the physical setting of the Mekong River Basin. Chapter 2 examines the evolution of the Mekong legal regime from its inception in 1957 to the present. Chapter 3 chronicles the evolution of Article 5(A)[15] (the "tributary rule") of the 1995 Mekong Agreement and examines this

15 Article 5(A) of the 1995 Mekong Agreement provides that: "On tributaries of the Mekong River, including Tonle Sap, intra-basin uses and inter-basin diversions shall be subject [only] to *notification* to the [Mekong River Commission] Joint Committee." (emphasis added).

salient provision of the treaty through the lens of selected international and interstate legal regimes, emphasizing the disparate treatment of tributaries within them. Chapter 3 evidences the shift doctrinally away from the terms of reference to the parties' interdependencies in the Mekong Committee era (1957–1978) to a unilateral approach to development manifested in Article 5(A) in the 1995 Mekong Agreement. Article 5(A) is transformative and is already having major effects, not only at state levels but at local levels, as well as raising concerns about whether sustainable development is a concept at odds with itself on Mekong tributaries. Drawing on the continuing dispute between Vietnam and Cambodia over hydropower development on the Sesan River at Yali Falls Dam[16] and the controversy over the "Thai water grid,"[17] Chapter 4 examines the extent to which the parties to the 1995 Mekong Agreement can effectively marginalize tributaries in the pursuit of national objectives and legally sidestep holistic water resources management and basin-wide sustainable water development envisioned in the cooperative spirit of the Agreement. In Chapter 5, a new term is proposed in the hydropolicy and legal literature, the "dissevered river basin," which is a generalized concept of a drainage basin used in a legal regime in which tributaries and/or headwaters are marginalized and/or excluded, and accordingly, prevents a holistic approach to water resources management (basin and sub-basin management under one umbrella agreement) and promotes hydrosovereignty. Chapter 6 reviews deficiencies in the Mekong legal regime that are related to and bear on the marginalization of tributaries, sustainable water development, and holistic water resources management. The Mekong legal regime is reviewed through the lens of the most comprehensive and influential articulation of customary law and obligations, the 1997 United Nations (UN) Convention on the Law of the Non-Navigational Uses of International Watercourses (UN Watercourses Convention),[18] which entered into force in August 2014. Ironically, it was a LMRB state, Vietnam, which was the 35th state to become a party to the UN

16 *See generally,* Michael Lerner, "Dangerous Waters: Violations of International Law and Hydropower Development Along the Sesan River" (unpublished manuscript, Oxfam, Phnom Penh, Cambodia, December 2003). [hereinafter Dangerous Waters]; Eric Rutkow et al., "Down River: The Consequences of Vietnam's Se San River Dams on Life in Cambodia and Their Meaning in International Law" (NGO Forum on Cambodia, Phnom Penh, Cambodia, December 2005) [hereinafter Down River] for exposés on the impacts of hydropower development at Yali Falls dam on the Sesan River.

17 *See generally,* Molle and Floch (2007), *supra* note 14 for a discussion of the scale of the "Thai water grid."

18 21 May 1997, U.N. Doc. A/RES/51/869, *reprinted in* 36 ILM 700 (1997) [hereinafter UN Watercourses Convention].

Watercourses Convention, triggering the treaty's entry into force. In Chapter 7, consideration is given to the need for a paradigm shift in the Mekong legal regime. The proposal is advanced that Cambodia, the Lao PDR, and Thailand should now join the UN Watercourses Convention, thereby strengthening the relationship between the 1995 Mekong Agreement and customary international law, and demarginalizing Mekong tributaries. Through this approach, the author hopes to elucidate for policy makers options that will foster holistic management of water resources and basin-wide sustainable water development. This approach will provide a means of balancing environmental, economic, and social interests as the Mekong countries transform from developing to developed states.

Analysis of the complex and convoluted Mekong legal regime reveals that the Mekong, with its "marginalized"[19] and "pruned"[20] tributaries and "lopped headwaters,"[21] is the archetypal dissevered river basin: separated, divided into parts, and broken up by the ambiguous verbiage and semantic gymnastics in the 1995 Mekong Agreement. In a flurry and blizzard of undefined and ambiguous terms ("basin," "river basin," "basin level," "basin-wide," "river system," and "proposed use") used throughout the treaty, its text generalizes and redefines the concept of a river basin[22] in which tributaries are marginalized and headwaters are excluded, yet on paper this vaunted Agreement creates the legal fiction of a holistic water resources management paradigm by clinging fervently to the "imported"[23] concept of sustainable development. In effect, tributaries in the Mekong legal regime are managed by each state unilaterally or at best, bilaterally, through application of the limiting language embodied in Article 5(A), the tributary rule, which emphasizes hydrosovereignty. The modern trend and badge of responsibility under the UN Watercourses Convention are to provide a legal framework for the entire watercourse, defined as including its tributaries,[24] and to promote basin-wide governance. Accordingly, in a functional, basin-wide legal framework, rules governing tributaries,

---

19 Carrard, *supra* note 10.

20 Louis Lebel and Po Garden, "Deliberation, Negotiation and Scale in the Governance of Water Resources in the Mekong Region" in C. Pahl-Wostl et al. (eds.), *Adaptive and Integrated Water Management* (Springer, 2008) 216.

21 *Id.*

22 A river basin is defined as "the entire area drained by a river *and its tributaries*." R.L. Bates and J.A. Jackson (eds.), *Glossary of Geology* (American Geological Institute, 1980).

23 Bastien Affeltranger, Intra-basin conflict resolution in the Mekong basin: Is a reconciliation of water values possible? 54 *in* International workshop on the value of water: Different approaches in transboundary water management (2005).

24 McCaffrey, *supra* note 9, at 35.

especially transboundary ones which are international rivers, should comport with those for the mainstream. In the Mekong legal regime, the limited extent to which planned measures and megaprojects on tributaries are subject to interdependence and basin-wide governance remains problematic.

The book concludes with a critical appraisal of the conceptual framework of the Mekong legal regime. The author asserts that although the Mekong legal regime is a good institutional framework evidencing long-term interstate cooperation during periods of conflict, it requires a fundamental rethink about tributaries and thus, holistic water resources management. More than 20 years of experience has shown that there are aspects of the 1995 Mekong Agreement that should be strengthened in order to secure the environmental, economic, and social benefits it promises. The lack of substantive and procedural frameworks at the tributary or sub-basin level and lack of institutional capacity in the MRC portend an era of new conflict that could occur in the LMRB if the river system is not managed holistically and the proper channels are not followed. Coupled with the UN Watercourses Convention, the 1995 Mekong Agreement can provide a future pathway for effective holistic water resources management and sustainable water development in one of Southeast Asia's largest and most important river basins. But, as Vick so aptly notes, "[I]mplementing international cooperation in water use, management and protection is not an easy task. In the most favourable conditions, cooperation is the result of lengthy negotiations and unabated good will."[25]

---

25 Margaret J. Vick, "The Senegal River Basin: A Retrospective and Prospective Look at the Legal Regime" (2006) 46 *Nat. Resources J.* 211 (quoting Dante Caponera, "Patterns of Cooperation in International Water Law: Principles and Institutions" (1985) 25 *Nat. Resources J.* 563, 569).

CHAPTER 1

The Mekong River Basin

1.1 Physical Setting

Rising from snowmelt at its source more than 5,000 meters (m) above sea level in the Tanggula Shan in the Tibetan Himalayas, it is called the *Dza Chu* (Water of Stone).[26] Discovered only in 1994 by a Japanese-Chinese joint research party, the Mekong headwaters are located in Qinghai Province at an altitude of 5,224 m at the foot of the glacier that discharges into Lasagongma Creek on the northern side of Guosongmucha Mountain.[27] Gathering force, this meltwater flows turbulently through deep gorges in the Yunnan province of China, where it is called the *Lancang Jiang* (Turbulent River), a name aptly describing the physical condition of the river there, characterized by whitewater, rapids, and whirlpools.[28] Coursing across the frontier of Asia's Golden Triangle (Myanmar, the Lao PDR, and Thailand borders), through terrains of seductive beauty and geologic diversity, then veering into the Lao PDR near the boundary with Thailand, it is called the *Mae Nam Khong* (Mother of Waters), from which it derives its westernized name, the "Mekong."[29] After rushing mightily over Khone Falls and crossing into Cambodia, it is called the *Tonle Thom* (Great Water).[30] And finally, after flowing 4,880 kilometers (km) "in a northwest-southeast trending pan-shaped basin"[31] through six countries,[32] changing character significantly,

26 Mekong Sources, "Mekong—The Mother of Waters" (2004) 1 (on file with author) [hereinafter Mekong Sources]; David Phillips et al., "Trans-boundary Water Co-operation as a Tool for Conflict Prevention and Broader Benefit Sharing" (prepared for the Sweden Ministry for Foreign Affairs, 2006) 90; Ariel Dinar et al., *Bridges Over Water: Understanding Transboundary Water Conflict, Negotiation and Cooperation, World Scientific Series on Energy and Resource Economics* (World Scientific Publishing Co., 2007) vol. 3, 225.

27 Michel Peissel, *The Last Barbarians: The Discovery of the Source of the Mekong in Tibet* (Henry Holt and Company, 1997); Masayuki Kitamura, Discovery and First Descent of the Mekong Headwaters, 1 Japanese Alpine News at 1 (October 2001).

28 Mekong Sources, *supra* note 26, at 1.

29 *Id.*

30 *Id.*

31 Avijit Gupta (ed.), *Large Rivers: Geomorphology and Management* (John Wiley and Sons Ltd., 2007) 435.

32 A 234 km stretch of the Mekong forms the border between Myanmar (Burma) and the Lao PDR.

 | DOI 10.1163/9789004362604_003

and meandering lazily through the Vietnamese landscape before discharging into the South China Sea through a vast delta (55,000 km^2), it is called the *Cuu Long* (River of Nine Dragons).[33] A watercourse with many names, the Mekong is the major river artery of Southeast Asia (Figure 1.1)[34] and is both a contiguous (serving as a boundary between countries) and transboundary or successive river (crossing the territories of states). Transecting a basin covering about 795,000 km^2, the Mekong ranks as the world's 10th largest river in terms of runoff (475,000 million m^3) and is the region's longest river.[35] The river and its tributaries form the "Mekong River Basin," a river *system* which transcends geopolitical boundaries.[36]

From its earliest days, it has been the object of desire. At once inviting and forbidding, the Mekong River Basin has always been shrouded in mystique. If one were to search the world for a setting that would lend itself to impenetrable mystery, this largely undeveloped watershed would not disappoint. The Mekong region has only recently been seen by Westerners as something other than a staging ground for superpower conflict[37] and wars of "national liberation."[38] In fact, for decades war precluded any serious attempts to develop the basin, as several policy disagreements among the Mekong states led to political and military conflicts.[39] As the 20th century ended, the Mekong was "dominated by an institutional regime that stressed the rapid collection

33 Mekong Sources, *supra* note 26, at 1.

34 Avijit Gupta and S.C. Liew, "The Mekong from Satellite Imagery: A Quick Look at a Large River, (2007) 85 *Geomorphology* 259, 262.

35 Mekong Committee, "The Mekong Committee: A Historical Account (1957–1989)" (Mekong Committee Secretariat, Bangkok, Thailand, 1989) 3; Carmen Revenga et al., *Watersheds of the World: Ecological Value and Vulnerability* (World Resources Institute, 1998) 2–92; Karen Bakker, "The Politics of Hydropower: Developing the Mekong" (1999) 18 *Pol. Geo.* 209; Aaron T. Wolf, *Atlas of International Freshwater Agreements* (United Nations Environment Programme (UNEP), 2002) 69; Evan Gottesman, *Cambodia After the Khmer Rouge: Inside the Politics of Nation Building* (Yale University Press, 2003) 12; Dinar et al., *supra* note 26, at 225; Gupta, *supra* note 31, at 3.

36 Anond Snidvongs and Seng-Keh Teng, "Mekong River" (GIWA Regional Assessment 55, Global International Waters Assessment, 2006) 16.

37 Arun P. Elhance, *Hydro-Politics in the 3rd World: Conflict and Cooperation in International River Basins* (United States Institute of Peace Press, 1999) 191.

38 *Khruschev Report on Moscow Conference of Representatives of Communist and Workers Parties* (John Fitzgerald Kennedy Library, Folder 4, President's Office Files, Box 126(a), 1961) 23 [unpublished, on file with author; hereinafter Khruschev Report].

39 Mekong Sources, *supra* note 26, at 1.

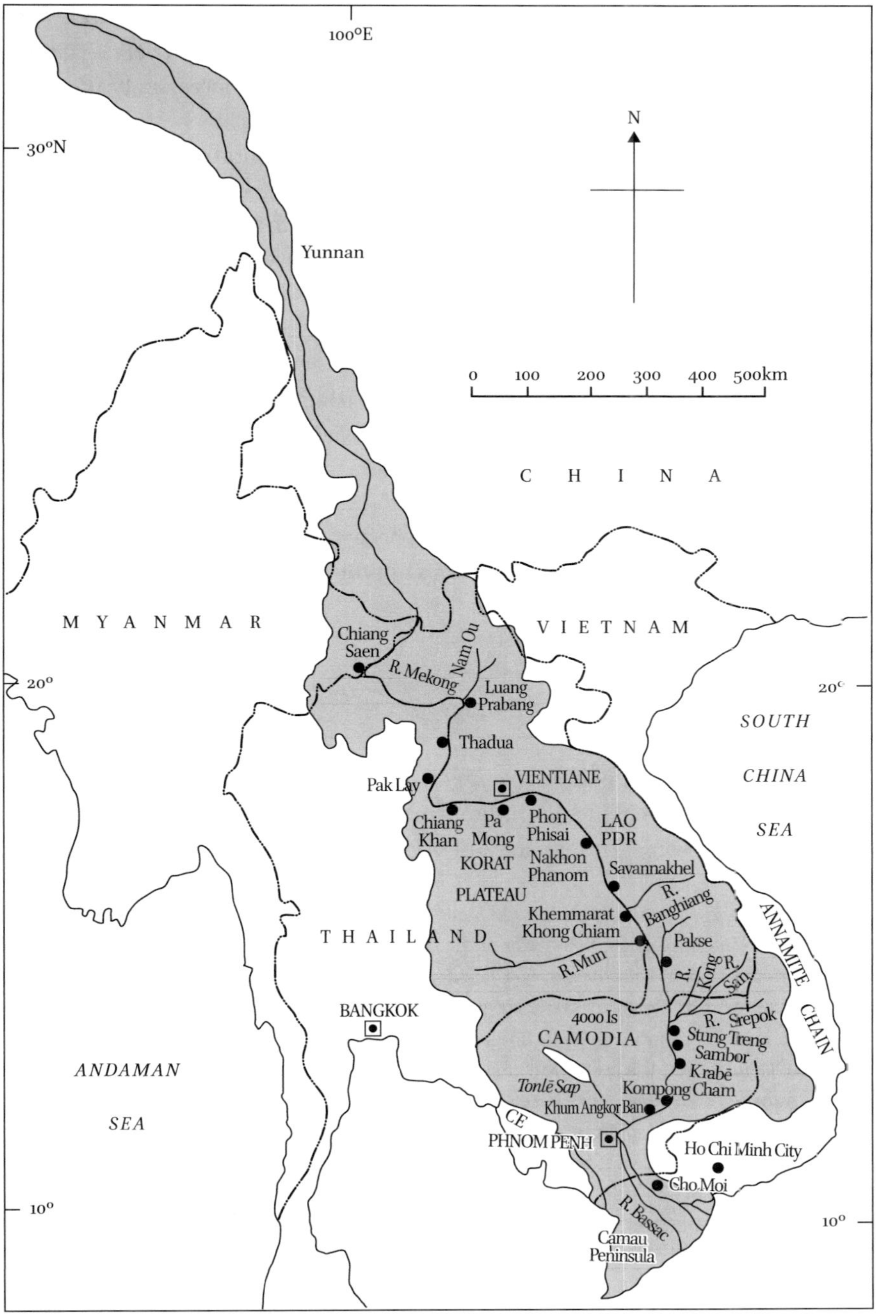

FIGURE 1.1 *The Mekong River Basin.* CE *northwest of Phnom Penh refers to the location of the Cardamon and Elephant ranges.*

SOURCE: GUPTA AND LIEW (2007). REPRINTED WITH PERMISSION FROM ELSEVIER.

of hydrologic data in order to convert it into a 'working' river."[40] Today the Mekong is recognized for its immense potential for hydropower, irrigation, navigation, and fishery development.[41] Contested and increasingly the focus of international attention, the Mekong remains a figurative battlefield, albeit one where economic development engages environmentalism.

The Mekong flows through two distinct regions, the Upper Mekong River Basin (UMRB) and LMRB which are divided into six geomorphic zones: the Lancang River Basin (UMRB in China), the Northern Highlands, the Korat-Sakon Plateau, the Eastern Highlands, the Southern Uplands, and the Lowlands.[42] The LMRB begins at the confluence of the Nam Ruk tributary and the mainstream (at km 2,382) where the Golden Triangle is located.[43] The LMRB covers about 70 percent of the entire basin.[44] When it enters the LMRB in the Golden Triangle, the river decreases in altitude by almost 4,500 m.[45] The Mekong drainage network is extremely complex and affects the lives of some 70 million people whose "different ethnic and cultural identities … form a complex and often antagonistic pattern of diversity challenging any unifying role the river might play"[46] and who depend on the river and its tributaries for subsistence.[47]

1.2 Geology and Geomorphology

The iconic Hindu Kush-Himalaya region is the source of a host of the world's great rivers including the Brahmaputra, Ganges, Indus, Irrawaddy, Yangtze, Yellow, and Mekong. The geology of the Mekong River Basin (Figure 1.2) is

40 R. White, *The Organic Machine: The Remaking of the Columbia River* (Hill and Wang, 1995); MRC, "Mekong News" (Newsletter of the MRC, April-June 2002 (issue no. 2)) 1; Chris Sneddon and Coleen Fox, "Rethinking Transboundary Waters: A Critical Hydropolitics of the Mekong Basin" (2006) 25 *Political Geography* 181, 185.

41 Mikiyasu Nakayama, "Aspects Behind Differences in Two Agreements Adopted by Riparian Countries of the Lower Mekong River Basin" (1999) 1 *J. Comp. Policy Anal.* 293, 294.

42 Snidvongs and Teng, *supra* note 36, at 17–19.

43 Bantita Pichyakorn, "International Watercourses Law: The Experience of the Mekong River Basin" in Surya P. Subedi (ed.), *International Watercourses Law for the 21st Century: The Case of the River Ganges Basin* (Ashgate Publishing, 2005) 167.

44 Snidvongs and Teng, *supra* note 36, at 17.

45 MRC, Overview of the Hydrology of the Mekong Basin 5 (2005).

46 D.E. Weatherbee, "Cooperation and Conflict in the Mekong River Basin" (1997) 20 *Studies in Conflicts and Terrorism* 167, 168.

47 Lazarus, Kate (ed.) et al., "Lancang-Mekong: A River of Controversy" (International Rivers Network [hereinafter IRN] and Mekong Watch, 2003); Snidvongs and Teng, *supra* note 36, at 22.

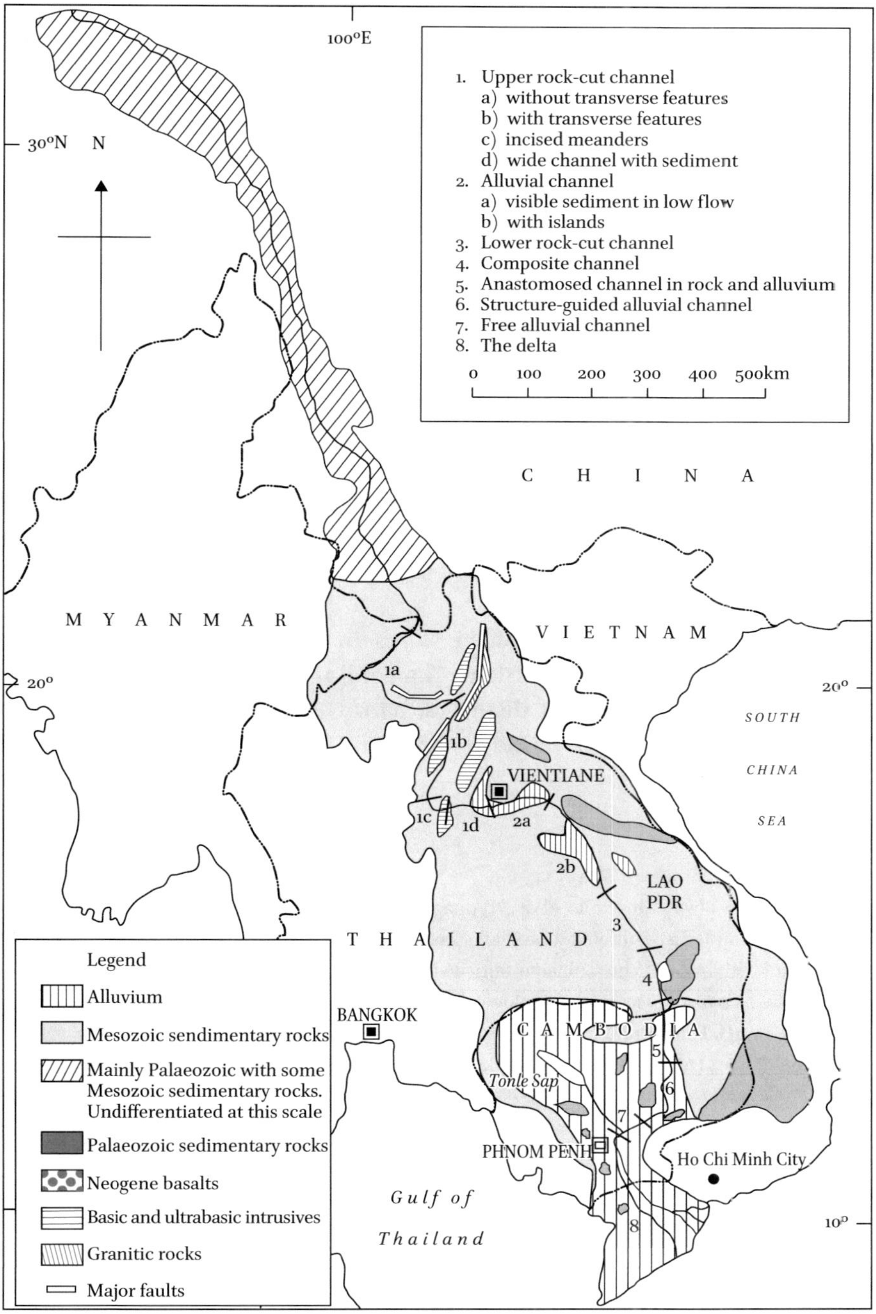

FIGURE 1.2 *Generalized geology and the Eight River Units identified by Gupta and Liew (2007) in the Lower Mekong River Basin.*
REPRINTED WITH PERMISSION FROM JOHN WILEY AND SONS, LTD.

complex and remains somewhat generalized in the geoscience literature; however, Gupta provides an excellent "snippet" of the heterogeneous basin architecture.[48] It is conjectured that the Mekong valley owes its genesis to extrusion tectonics, specifically to the collision of the Indian and Eurasian plates and the attendant Himalayan Orogen and rising of the Tibetan Plateau.[49]

The UMRB (China) is characterized by deep, narrow gorges of the Lancang River Basin which dissect Paleozoic and Mesozoic granites and sedimentary rocks.[50] In Yunnan Province, China, the Mekong flows turbulently through a narrow rock valley with precipitous slopes.[51] The Lancang Jiang contributes 16 percent of the Mekong's flow and half of the sediment load.[52] Satellite imagery indicates a channel partly filled with sediment.[53]

From the Yunnan Province southward to the Lao PDR, the Mekong basin consists of "granitic rocks, folded Paleozoic sedimentary and metamorphic rocks (sandstone, limestone, chert, slate, schist), Mesozoic sedimentary deposits (sandstone, limestone, evaporites), and local volcanic exposures."[54] An ancient suture zone and its attendant basic and ultrabasic rocks also occur in this area.[55] Here in the deep valley of the river, the Mekong tracks several large faults.[56]

In the mountainous region of the Northern Highlands (southern Yunnan Province, eastern Myanmar, northern Thailand, and the Lao PDR), the Mekong landscape "is a collection of dissected, narrow-crested, near parallel ridges separated by deep valleys."[57] Ridge elevations in the highlands rise 1,500 to

---

48 Gupta, *supra* note 31, at 437–439.

49 *See generally,* P. Tapponier et al., "Propagating Extrusion Tectonics in Asia, New Insights from Simple Experiments with Plasticine" (1982) 10 *Geology* 611.); P. Tapponier et al., "On the Mechanics of the Collision between India and Asia" in M.P. Coward and A.C. Ries (eds.), *Collision Tectonics* (Geological Survey of London Special Publication 19, 1986) 115–157; G. Peltzer and P. Tapponier, "Formation and Evolution of Strike-Slip Faults, Rifts and Basins During the India-Asia Collision: An Experimental Approach" (1988) 93 *Journal of Geophysical Research* 15085–15117, describing the tectonic events that developed the basin architecture in the Mekong region.

50 Snidvongs and Teng, *supra* note 36, at 18.

51 Gupta, *supra* note 31, at 437.

52 MRC, *supra* note 45, at 1.

53 Gupta, *supra* note 31, at 437.

54 *Id.*

55 *Id.*

56 *Id.*

57 Snidvongs and Teng, *supra* note 36, at 18; Gupta, *supra* note 31, at 437.

2,000 m with corresponding valleys of 1,000 m.[58] Limestone cliffs overlook the mainstream and its major tributaries.[59] The Annamite Chain of the Eastern Highlands, with elevations up to 2,000 m, forms the eastern divide between Vietnam and the Lao PDR or Cambodia.[60] Also in this geomorphic region are the Central Highlands, a vast upland plateau situated between the Cambodian border and the South China Sea. Major tributaries (Se Bang Fai, Se Bang Hiang, Sekong, Sesan, and Srepok) draining the Highlands have eroded the mountains.[61] Tributaries in this region contribute more than two-fifths of the total water volume in the Mekong basin.[62] The western divide is marked by the low ridges of the driest region of the basin, the Korat-Sakon Plateau (200–500 m), which is drained by the Mun and Chi rivers and their tributaries.[63] Farther south in the upper LMRB near Vientiane, Lao PDR, Quaternary alluvium drapes older rocks.[64] Downstream from this alluvium, Mesozoic sedimentary rocks (sandstones and evaporites) crop out forming numerous "rock ribs" in the Mekong channel. Near the Lao PDR-Cambodian border, volcanic rocks (Neogene basalts) are exposed.[65]

Turning into Cambodia, the Mekong courses over "an alluvium of variable thickness" accentuated with local exposures of Triassic sedimentary rocks and Neogene volcanic rocks (basalt) near the river.[66] These rocks occur on the highlands that form the divides.[67] Drainage from the western embayment south of the Korat-Sakon Plateau collects in the most conspicuous feature on the Mekong landscape, the Tonle Sap (Great Lake).[68] Connected to the mainstream by the Tonle Sap River, which drains the Southern Uplands (Cardamon and Elephant ranges in southeastern Cambodia), the Tonle Sap in the Lowlands is the largest lake in Southeast Asia and seasonally reverses flow

---

58 Gupta, *supra* note 31, at 437.

59 *Id.*

60 Snidvongs and Teng, *supra* note 36, at 18; Gupta, *supra* note 31, at 437–439.

61 *Id.*

62 P. Hirsch and G. Cheong, "Natural Resource Management in the Mekong River Basin: Perspectives for Australian Development Cooperation" (Final Overview Report to AusAID, University of Sydney, 1996) 22.

63 Snidvongs and Teng, *supra* note 36, at 18; Gupta, *supra* note 31, at 439.

64 Gupta, *supra* note 31, at 437.

65 G. Choubert and A. Faure-Muret, *Geological Atlas of the World* (UNESCO, 1976); Economic and Social Commission for Asia and the Pacific (ESCAP), "Atlas of Mineral Resources of the ESCAP Region" (Explanatory Brochure, United Nations, 1990) 19.

66 Gupta, *supra* note 31, at 437.

67 *Id.*

68 *Id.* at 439.

in response to the Mekong's flood pulse.[69] In the southernmost Lowlands of the LMRB, the river flows on thick alluvium in its lowermost reaches in Vietnam's "Mekong delta."[70]

Structurally, grabens and half-grabens have formed Tertiary basins in Thailand.[71] According to Hutchinson, in the lowlands of Cambodia and Vietnam, alluvial fill has covered and obscured the pre-existing structural architecture (Tonle Sap graben).[72] Gupta posits that it is likely that the river "had to adjust to such basins across its drainage area."[73] In the Lowlands, the present Mekong delta "postdates the wide shifting of the coastline during the Pleistocene."[74] The geologic evolution of the Mekong basin still remains virtually unknown, a befitting tribute to the mysterious river that flows through it endlessly to the South China Sea.

For a large river, the Mekong is "remarkably structure-guided," flowing for almost 80 percent (3,000 km) of its length through a narrow valley underlain with rock.[75] Over the next 1,000 km, steep mountainous slopes dominate the Mekong landscape as the river courses over rocks and alluvium.[76] Alluvium underlies only the last 600 km of the river's reach as it transects "a 500 km wide lowland that converges to a large delta."[77] In the UMRB and the upper to mid- LMRB, Mekong stream channels are characterized by steep banks. LMRB stream beds in Cambodia and the delta are composed of thick silt and sand. Stream gradients are highest in the first 2,500 km of the Mekong's course but vary in the LMRB dependent upon the uniformity of the erodible substrate.[78] Sediment from erosion of incised valleys is deposited downstream along floodplains.

Based on satellite imagery, topographic maps, charts from the MRC's hydrographic atlas, water data from the MRC's Lower Mekong Hydrologic Yearbooks, and field reconnaissance data, Gupta and Liew analyzed the river's morphology framed in the context of its little-known evolution.[79] They divided the lower

69 Snidvongs and Teng, *supra* note 36, at 18–19; Gupta, *supra* note 31, at 439.

70 Snidvongs and Teng, *supra* note 36, at 19; Gupta, *supra* note 31, at 437.

71 E.A. Lorenzetti et al., "Structural Styles in Rift Basins: Interpretation Methodology and Examples from Southeast Asia" (1994) 78 *AAPG Bulletin* 1152.

72 C.S. Hutchinson, *Geological Evolution of Southeast Asia*, Oxford Monographs on Geology and Geophysics no. 13. (Clarendon Press, 1989) 368.

73 Gupta, *supra* note 31, at 437.

74 *Id.*

75 *Id.*

76 *Id.*

77 *Id.*

78 *Id.* at 443.

79 Gupta and Liew, *supra* note 34, at 264–270.

Mekong from the China border to the South China Sea into eight segments or "River Units": (1) Upper rock-cut channel (Subunits 1a–without transverse features (e.g., rapids or knickpoints), 1b–with transverse features, 1c–incised meanders, and 1d–wide channel with sediment); (2) alluvial channel (Subunits 2a–visible sediment in low flow and 2b–with islands); (3) lower rock-cut channel; (4) composite channel; (5) anastomosed channel in rock and alluvium; (6) structure-guided alluvial channel; (7) free alluvial channel; and (8) the delta (Figure 1.2).[80] Similarly, the MRC divided the Mekong into six reaches based on hydrological regime, physiography, land use, and existing, planned and potential resource developments: Reach 1: "Yunnan Component"—Lancang Jiang or Upper Mekong in China; Reach 2: Chiang Saen to Vientiane and Nongkhai; Reach 3: Vientiane and Nongkhai to Pakse; Reach 4: Pakse to Kratie; Reach 5: Kratie to Phnom Penh: and Reach 6: Phnom Penh to the South China Sea.[81]

The Mekong (Lancang Jiang-Reach 1) in Yunnan Province, China, is structure-guided and flows turbulently through steep rock gorges.[82] Small tributaries conjoin with the mainstream, draining very steep small watersheds.[83] This "Yunnan Component" is the center of considerable political attention because it contains the Mekong's headwaters and, accordingly, "plays a central role in the low-flow hydrology of the lower mainstream."[84]

From southern Yunnan Province to Vientiane (River Unit 1—Subunits 1a through 1d, Reach 2), the Mekong flows in the upper rock-cut channel, but the morphology changes several times over its 900-km course from the China border.[85] River Subunit 1a is a rocky conduit with a number of straight reaches that coincide with faults.[86] This Subunit contains few transverse features, such as rapids, which characterize River Subunit 1b in which the Mekong flows on rock with banks in inset sand and silt.[87] At the terminus of Subunit 1b, according to Gupta and Liew, "without any apparent reason, the Mekong suddenly changes from a south-flowing river to an east-flowing one" with incised meanders that end at Chiang Khan where the river courses north, widens and straightens in River Subunit 1d.[88] On the Thai side of Reach 2, there is an elaborate scheme, the so-called "Thai water grid,"[89] to divert water out of Lao PDR and

80 Gupta, *supra* note 31, at 438; Gupta and Liew, *supra* note 34, at 264–270.

81 MRC, *supra* note 45, at 7–9.

82 Gupta, *supra* note 31, at 443.

83 *Id.*

84 MRC, *supra* note 45, at 8.

85 *Id.*; Gupta, *supra* note 31, at 443.

86 Gupta, *supra* note 31, at 444.

87 Gupta and Liew, *supra* note 34, at 265.

88 *Id.* at 266.

89 MRC, *supra* note 45, at 8.

Cambodian tributaries into Thailand. Reach 2 hydrology is dominated by the Yunnan Component.[90]

The Mekong flows in an alluvial channel rather than on rock over the next 500 km.[91] River Units 2 (Subunits 2a and 2b) and 3 are contained in the MRC's Reach 3. The boundary between Reach 2 and Reach 3 evidences significant change in Mekong hydrology.[92] It is at this junction within Reach 3 where the mainstream becomes increasingly influenced by the large left bank tributaries in the Lao PDR, such as the Se Bang Fai and Se Bang Hiang and the Mun-Chi river system from the right bank.[93] The MRC describes the former as "in an accelerating phase of development in terms of water demand for agriculture and hydropower" and the latter as "highly developed low-relief, agricultural basins with low runoff potential and significant reservoir storage for dry season irrigation."[94] Subunit 2a is a wide and shallow meandering channel with sediment.[95] Point bars, visible in low flow, store sediment on vertical banks.[96] Downstream of this Subunit, the Mekong continues another 400 km as a near-straight alluvial channel with solitary bars and islands (Subunit 2b).[97] Downstream of Subunit 2a, the Mekong returns to a lower rock cut channel in River Unit 3, which continues for 200 km with several instream exposed rock islands and close hills along the banks.[98] The Mun River drains the Korat Plateau and enters the mainstream near the end of Unit 3.[99]

River Units 4, 5 and 6 (MRC's Reach 4) comprise the Mekong composite channel, anastomosed channel in rock and alluvium and structure-guided alluvial channel, respectively.[100] In the composite channel (Unit 4), the Mekong flows over rocks for 150 km with high alluvial banks.[101] In the next 200 km, the Mekong "alternates between near-straight and anastomosing channels" (Unit 5), the latter flowing over Mesozoic basalt, forming a zone of waterfalls and rapids.[102] It is in Unit 5 where arguably the most important tributaries, the

90 *Id.*
91 Gupta, *supra* note 31, at 445.
92 *Id.*
93 MRC, *supra* note 45, at 8.
94 *Id.*
95 Gupta, *supra* note 31, at 445.
96 Gupta and Liew, *supra* note 34, at 266.
97 *Id.* at 267.
98 *Id.*
99 *Id.* at 268.
100 *Id.* at 268–269.
101 Gupta, *supra* note 31, at 447.
102 *Id.*

Sekong, Srepok, and Sesan Rivers, which drain the Annamite chain, contribute over 25 percent of the mean annual flow to the Mekong mainstream near Stung Treng.[103] These three major tributaries are a key element in the hydrology of the flow reversal of the Tonle Sap.[104] Approaching the 500 km wide Cambodian plain, River Unit 6 of the Mekong remains "rock-controlled" as it traverses the LMRB adjacent to Triassic sedimentary rocks and Neogene basalts.[105]

As the structure-guided alluvial channel (Unit 6) transitions into the free alluvial channel (Unit 7, MRC's Reach 5), the Mekong meanders across a wide floodplain marked with abandoned channels and meander scroll.[106] At this juncture, 95 percent of the total flow has entered the river system and the LMRB is a large wetland.[107] The Bassac, the Mekong's first distributary, signals the gateway to the delta (Unit 8, MRC's Reach 6) at Phnom Penh as the river makes its way to its terminus in the South China Sea.[108]

1.3 Climate

Climate ranges from cold temperate and tundra in the UMRB to monsoonal in the LMRB.[109] In the UMRB, the climate is generally cold and lacks a definable summer season, and mountains in the Tibetan Plateau are snow-capped year-round.[110] Clearly defined seasons and rainfall up to 1,700 millimeters (mm) occur farther south in the subtropical climate of Yunnan Province.[111] Tropical monsoonal conditions predominate in the LMRB with onset of the southwest (SW) monsoon (rainy season), which brings 75 percent of the total rainfall beginning in mid-May and continuing to early October.[112] The SW monsoon

103 MRC, *supra* note 45, at 9; Gupta, *supra* note 31, at 447.

104 MRC, *supra* note 45, at 9.

105 Gupta, *supra* note 31, at 447.

106 MRC, *supra* note 45, at 9; Gupta, *supra* note 31, at 448.

107 Gupta, *supra* note 31, at 448.

108 *Id.*

109 Snidvongs and Teng, *supra* note 36, at 19.

110 H. Liu, "Water Resources Simulation Model of the Lancang River (Yunnan Portion)" in D. He, G. Zhang, and K. Hsiang-te (eds.), *Towards Cooperative Utilization and Coordinated Management of International Rivers* (Science Press, 2001) 208–215.

111 T.H. Chu et al., "Water, Climate, Food and Environment in the Mekong Basin in Southeast Asia" (Final Report, International Water Management Institute [hereinafter IWMI], Mekong River Commission [hereinafter MRC] and Institute of Environmental Studies, 2003); Snidvongs and Teng, *supra* note 36, at 19.

112 Snidvongs and Teng, *supra* note 36, at 19.

results in a flood pulse that moves the water downstream. Flooding occurs in the rainy season usually during July through September.[113] The Mekong peaks in mid-August or early September in the upper LMRB and in mid-September or early October in the delta.[114] The northeast monsoon (dry season) begins in early November and continues through early March.[115] The hottest months in the LMRB correspond with the beginning of the dry season with average temperatures ranging from 30 to 38° C.[116] The lowest temperatures occur in the LMRB between November and February.[117]

Annual rainfall varies geographically in the region, increasing southward from 600 mm in the Lancang River Basin (UMRB) to 3,200 mm in the Eastern Highlands.[118] Average precipitation in the LMRB is 1,672 mm.[119] Rainfall decreases northward over China (1,000 mm) and the inherently dry Korat-Sakon Plateau.[120] Precipitation in the Annamite Chain and Central Highlands is some of the heaviest in the basin (2,000 to 3,200 mm).

1.4 Water Resources

The Mekong River Basin contains abundant surface water resources with a cumulative discharge of approximately 475,000 million m^3, 78,800 million m^3 of which occurs in the dry season.[121] Dry season discharges are 4 to 7 times lower than their rainy season counterparts.[122] The LMRB contains approximately 77 percent, or about 612,150 km^2, of the entire basin area.[123] In the UMRB, China contributes 16 percent of the flow to the Mekong and 21 percent of the basin area (165,000 km^2 or 2 percent of the country).[124] Myanmar provides a scant 2 percent of the flow and 3 percent of basin area (4 percent of

113 Chu et al., *supra* note 111; Snidvongs and Teng, *supra* note 36, at 20.
114 Snidvongs and Teng, *supra* note 36, at 20.
115 *Id.*
116 *Id.*
117 *Id.*
118 Snidvongs and Teng, *supra* note 36, at 19.
119 Gupta, *supra* note 31, at 439.
120 *Id.*
121 Snidvongs and Teng, *supra* note 36, at 20.
122 *Id.*
123 *Id.* at 11; MRC, "Water Utilization Program Preparation Project: Final Report", MKG/R.98/013 (1998); Gupta, *supra* note 31, at 440.
124 Snidvongs and Teng, *supra* note 36, at 20.

the country or 24,000 km^2), the lowest of any riparian state.[125] In the Lao PDR, the Mekong drains almost the entire country (202,000 km^2 or 97 percent), contributing the lion's share of the flow (35 percent) in one quarter of the drainage basin.[126] Thailand's basin area is 23 percent (36 percent of the country or 184,000 km^2) and its flow volumetric contribution is 17 percent, only 1 percent higher than China's flow.[127] An impressive 86 percent of Cambodia (155,000 km^2) lies within the LMRB contributing 20 percent to the basin area but only 19 percent of the total Mekong flow.[128] Vietnam, despite the presence of the sprawling Mekong delta, contributes only 8 percent of the basin area (65,000 km^2 or 20 percent of the country) and 11 percent of the LMRB flow.[129] Average flow contributions are: (1) China—2,410 cubic meters per second (m^3/sec), (2) Myanmar-300 m^3/sec, (3) Lao PDR—5,270 m^3/sec, (4) Thailand—2,560 m^3/sec, (5) Cambodia—2,860 m^3/sec, and (6) Vietnam—1,660 m^3/sec, for a total average flow of 15,060 m^3/sec.[130]

Water availability and withdrawals in the Mekong River Basin are provided in Table 1.1.[131] Surface water availability (internally annual renewable resources) varies widely across the basin. The Lao PDR has the largest per capita availability and Yunnan Province (China) has the lowest (Table 1.1). Vietnam has the highest per capita withdrawals followed by Thailand and Yunnan Province (Table 1.1). Urbanization, irrigation, and industrialization are increasing the water demand in the basin, especially in Thailand and Vietnam. Vietnam has the highest ratio of withdrawals to availability (Table 1.1). Irrigated agriculture is the largest user of water and accounts for 94 percent of the total withdrawals in Cambodia, 82 percent in the Lao PRD, 91 percent in Thailand, and 86 percent in Vietnam.[132] Bangkok and other urban centers depend primarily on the Mekong and its tributaries for their water supply.

---

125 *Id.*

126 *Id.*

127 *Id.*

128 *Id.*

129 *Id.*

130 MRC, *The Mekong Programme*, "Table 1: Approximate Distribution of MRB Water Resources by Country" <http://archive.iwlearn.net/www.MRCmekong.org/www.MRCmekong.org/mekong_program_ceo.htm> (accessed 8 June 2017).

131 ESCAP, "Sources and Nature of Water Quality Problems in Asia and the Pacific", ST/ESCAP/1874 (United Nations, 1998); Claudia Ringler, *Optimal Allocation and Use of Water Resources in the Mekong River Basin: Multi-Country and Intersectoral Analyses* (Development Economics and Policy Series, Peter Lang, 2001) vol. 20, 46.

132 Chu et al., *supra* note 111.

TABLE 1.1 *Water resources availability and withdrawals in the Mekong River Basin, 1995*

	Availability		**Withdrawals**		**Withdrawal share of availability**
	(km^3/yr)	*($m^3/cap/yr$)*	*(km^3/yr)*	*($m^3/cap/yr$)*	(%)
Cambodia	88	8,585	1	98	1
Lao PDR	270	55,305	1	205	<1
Myanmar	606	13,024	4	86	<1
Thailand	210	3,559	33	559	16
PRC	2,812	2,292	500	407	18
Viet Nam	318	4,479	65	915	20

Note: Availability refers to annual internally renewable water resources.
SOURCE: CLAUDIA RINGLER, "OPTIMAL WATER ALLOCATION IN THE MEKONG RIVER BASIN" (ZEF DISCUSSION PAPERS ON DEVELOPMENT POLICY NO. 38, CENTER FOR DEVELOPMENT RESEARCH, BONN, GERMANY, MAY 2001).

Chea accurately predicted that total water supply to Phnom Penh and Vientiane would increase from 155,000 m^3/day in 1993, to 272,000 m^3/day by 2010.[133] Industrial and agricultural water uses in Thailand and Vietnam are approaching 20 percent of total annual internally renewable resources.[134] Ringler estimates that domestic and industrial water use will increase more than two-fold from 1.89 billion m^3 in 1990 to 4.1 billion m^3 in 2020.[135] Continued water use at this magnitude will significantly alter the flow regimes of the Mekong and its tributaries from which, according to Snidvongs and Teng, "the majority of this demand will be met … with groundwater supplying a smaller proportion."[136]

Preliminary studies by the MRC in the LMRB indicate the potential for extensive groundwater resources of 60 million m^3/day, but these have not been

133 V. Chea, "Water Supply in Phnom Penh" in UN Economic Commission for Asia and the Pacific, *Towards Efficient Water Use in Urban Areas in Asia and the Pacific*, (United Nations, 1998) ST/ESCAP/1874.

134 Snidvongs and Teng, *supra* note 36, at 28.

135 Ringler, *supra* note 131, at 58.

136 Snidvongs and Teng, *supra* note 36, at 27.

adequately assessed.[137] Highly recharged aquifers of recent alluvium replenished through rainwater seepage flank the mainstream in Thailand and the Mekong delta.[138] Groundwater moving through these unconsolidated sediments issues from seeps and springs in the stream valleys and can serve as a source of stream discharge during the dry season. Since 1980, more than 6,000 groundwater wells have been drilled in these aquifers and other parts of the LMRB.[139] Already, over extraction of groundwater for agriculture has affected groundwater levels in the Vietnamese highlands.[140] From 1997 to 2000, extraction of groundwater for water supply and agriculture in Cambodia increased from 120,000 m^3/day to 290,000 m^3/day.[141] No significant lowering of the water table has occurred in the LMRB.[142]

A water quality monitoring program, now known as the MRC Water Quality Monitoring Network, was established in the LMRB in 1985.[143] Water quality in the Mekong is generally good, especially in the UMRB.[144] However, population growth, irrigation, and industrialization are causing significant concern about water quality in the LMRB, where major water quality problems have been found in some locales. Indeed, the LMRB illustrates the complexity of managing transboundary rivers because the use of water in one state can impact downstream states. Major land uses in the Mekong basin are agricultural (75 percent), forestry (17 percent) and residential, commercial, and industrial (0.1 percent), and Mekong water quality reflects the landscape

---

137 MRC, Mekong River Basin diagnostic study—Final report (1997); Chu et al., *supra* note 111; Several transboundary aquifers have been identified within the Mekong River Basin, namely the Lancang River Downstream Aquifer (China and Myanmar), the Mekong River midstream aquifer (Thailand, Laos and Vietnam), Khorat Plateau aquifer (Thailand and Laos), and the Mekong River delta aquifer (Cambodia and Vietnam). See Remy Kinna and Alistair Rieu-Clarke, "The Governance Regime of the Mekong River Basin" (2017) 2 *Int'l Water L.* 1–84, at 36, *citing* K. Ha, et al., *Current Status and Issue of Groundwater in the Mekong River Basin* (Bangkok: KIGAM, CCOP and UNESCO 2015), at 50.

138 Snidvongs and Teng, *supra* note 36, at 21.

139 *Id.*

140 *Id.* at 29.

141 Chu et al., *supra* note 111.

142 *Id.*

143 Deborah F. Shmueli, "Water Quality in International River Basins" (1999) 18 *Political Geography* 437, 451.

144 I.C. Campbell, "Integrated Water Resources Management for the Mekong River Basin" in A.K. Biswas et al. (eds.) *Integrated Water Resources Management in South and South-East Asia* (Oxford University Press, 2005) 353.

through which the river flows.[145] Potential transboundary water quality issues include sedimentation, water quality degradation, and water quantity alteration. The major sources of pollution are agricultural wastes, sediment, domestic sewage, and industrial wastes.[146] Organic wastes, bacteria, nutrients, heavy metals, pesticides, fertilizers, and other toxic substances are beginning to cause water quality problems in the LMRB.[147] The Mekong's flow in the dry season is "sometimes insufficient to flush out its sluggish waters."[148] The ammonia content in Mekong tributaries has significantly increased since 1993, reflecting the basin's increased use of fertilizers in agriculture.[149] Shmueli observed that "[p]ollution in Mekong tributaries is also *increasing at a steady rate*."[150]

Water quality issues in the LMRB are as diverse as the physical terrains through which the Mekong and its tributaries flow. Cambodia's concerns obviously focus around the Tonle Sap, "identified as the environmentally most critical water source."[151] Siltation from deforestation and turbidity from Mekong silt loads are real threats to lake storage capacity, fish populations, ecology, and thus the national economy.[152] Low dissolved oxygen and high organic pollutant loads from sewage in the dry season also plague the lake and Tonle Sap River, the vital inlet and outlet to the Mekong mainstem.[153]

In the Lao PDR, a state seeking to emerge from grinding rural poverty, wastewater from towns and industry is a source of water degradation.[154] Water hardness also occurs in tributaries, such as the Se Bang Fai, which flow through limestone terrains.[155] Sedimentation attributed to deforestation is a problem in several tributaries, draining the highlands during the rainy season.[156] Overall, due to abundant water resources and low population density, water quality in the Lao PDR remains good.[157]

145 Wanpaen Wirojanagud and Phornphimon Inthong, "Trans-boundary water quality in the Mekong River" (June 2008) *Water21* 59.
146 *Id.*
147 *Id.*
148 Shmueli, *supra* note 143, at 448.
149 Wirojanagud and Inthong, *supra* note 145, at 60.
150 Shmueli, *supra* note 143, at 450 (emphasis added).
151 Ringler, *supra* note 131, at 48.
152 *Id.*
153 *Id.*
154 *Id.* at 47.
155 *Id.*
156 *Id.* at 47–48.
157 *Id.* at 47.

Continued economic growth in Thailand's western-like economy has not come without an environmental price. Agricultural-industrial development in the drought-prone Northeast, or "Isan," through large-scale interbasin and sub-basin transfers of Mekong waters has caused substantial water quality problems.[158] Domestic wastewater is a major pollutant and has resulted in attendant low biochemical oxygen demand and pathogens.[159] Pulp and paper mills, textile factories, sugar mills, distilleries, and tapioca farms have contributed industrial pollutants to the river, as have animal (pigs and ducks) farms and aquaculture ponds.[160] The Mun River Basin suffers from hypersalinity, which exacerbates water quality problems for agribusiness and water supply on the Korat-Sakon Plateau where "salinization of farm land is increased by runoff from rock salt mining and application of irrigation waters upstream from dams."[161] Other pollutants include agrichemicals from runoff and nitrogen fertilizer used in rice paddy production.[162]

Deforestation and agriculture have caused erosion and sedimentation in the Mekong and its tributaries in Thailand. The MRC Water Quality Monitoring Network indicates that total suspended solids are higher in the rainy season and lower in the dry season.[163] Total suspended solids increase upstream in the Mekong in Thailand indicating that erosion is taking place upstream.[164] Wirojanagud and Inthong note that Mekong tributaries (the Mun and Songkhram) are not the source of the total suspended solids in the mainstem and "data about TSS [total suspended solids] content within the major tributaries … would be likely to find higher levels than at the outlet of these tributaries" reflecting that "sedimentation takes place in the river itself, and is not distributed across boundaries."[165]

In Vietnam, the main threats to water quality are domestic and industrial wastes.[166] Deforestation, agriculture, and urban stormwater are also major factors affecting water quality.[167] Perhaps Vietnam's most pressing water quality

158 Shmueli, *supra* note 143, at 450; Ringler, *supra* note 131, at 48.
159 Ringler, *supra* note 131, at 48.
160 *Id.*
161 Shmueli, *supra* note 143, at 448.
162 *Id.*
163 Wirojanagud and Inthong, *supra* note 145, at 60.
164 *Id.*
165 *Id.*
166 Ringler, *supra* note 131, at 49.
167 *Id.*

issue is the protection of the Mekong delta from saltwater intrusion.[168] Indeed, this was Vietnam's banner issue during the negotiation of the 1995 Mekong Agreement. Maintenance of dry-season flow to prevent seawater damage is of the essence for Vietnam's plans for increased rice paddy production in the delta to meet demands of a fast-growing population.

Although water quality in the Mekong and its tributaries is considered to be good, it could decline with increasing hydropower development, industrialization, use of pesticides and fertilizers, urban waste, sewage disposal, increased withdrawals for irrigation, and sedimentation caused by erosion from deforestation. As Shmueli observes "[t]he present pace of economic development, ... has already led to undesirable environmental effects."[169]

1.5 Mekong Tributaries

A network of more than 100 major and minor tributaries is a conspicuous feature of the LMRB landscape.[170] Indeed, the Mekong has a very long mainstream fed by numerous relatively short tributaries. Mekong tributaries differ significantly from the dendritic drainage pattern attendant to most large rivers.[171] Dendritic drainage is typically associated with the gentle slopes of basin architecture, consisting of homogeneous rocks in tectonically inactive terrains.[172] In marked contrast, it is the geologically heterogeneous and tectonically active terrains of the Mekong sub-basins that control the tributaries' courses and riverscapes, eroded by time's rivers to their current geomorphic forms.[173] Most of these tributaries rise in the LMRB.[174] Indeed, 77 percent of the Mekong River Basin's total drainage occurs in the LMRB, 35 percent of which originates in the Lao PDR.[175] Major tributaries contributing an average yearly discharge of 50 m^3/sec or greater to the Mekong are listed

168 Dinar et al., *supra* note 26, at 234.

169 Shmueli, *supra* note 143, at 448.

170 V.R. Pantulu, "The Mekong River System" in Bryan R. Davies and Keith F. Walker (eds.), *The Ecology of River Systems* (Springer, 1986) 706.

171 M. Clark et al., "Surface Uplift, Tectonics, and Erosion of Eastern Tibet from Large-Scale Drainage Patterns" (2004) 23 *Tectonics* TC1006, doi:10.1029/2002TC001402.

172 C. Twidale, "River Patterns and Their Meanings" (2004) 67(3–4) *Earth Science Reviews* 159.

173 *Id.;* S.K. Tandon and R. Sinha, "Geology of Large River Systems" in Gupta, *supra* note 31, at 7–28 (2007).

174 Dinar et al., *supra* note 26, at 227.

175 *Id.*

TABLE 1.2 *Major tributaries of the Lower Mekong River Basin* ($m^3s^{-1} = m^3/sec$)

Country	No. Tributary	Catchment area km²	Average yearly flow m^3s^{-1}
Lao PDR	1 NAM Tha	8,170[a]	140[a]
	2 Nam Ou	26,160[a]	430[a]
	3 Nam Suang	6,290[a]	100[b]
	4 Nam Khan	7,620[a]	130[b]
	5 Nam Ngum	17,600[b]	760[b]
	6 Nam Nhiep	4,690[b]	240[a]
	7 Nam Theum	14,700[b]	890[a]
	8 Se Bang Fai	9,470[a]	410[b]
	9 Se Ban Hieng	19,600[a]	530[b]
	10 Se Done	7,170[a]	230[a]
Thailand	11 Nam Mae Kok	10,800[b]	210[b]
	12 Nam Mae Ing	8,290[b]	110[b]
	13 Nam Loei	4,100[a]	50[b]
	14 Nam Songkhram	12,700[c]	300[b]
	15 Mun-Chi	117,000[b]	720[b]
Viet Nam + Kampuchea, Lao PDR	16 Se San (Se San + Se Kong + Sre Pok)	76,700[a]	2,900[a]
Kampuchea	17 Tonle Sap	84,400[d]	960[d]
	18 Prek Te	4,170[a]	85[a]
	19 Prek Chlong	5,750[a]	90[a]
	20 Prek Thnot	5,050[a]	60[a]

a Inventory of Promising Tributary Projects.
b Mekong Hydrological Yearbook; monthly flows of Mekong tributaries.
c Nam Songkhram Pre-feasibility Report.
d Estimated from total inflow to Great Lake (770 m^3s^{-1} for catchment of 67,610 km²); Great Lac due Cambodge, Sédimentation et Hydrologie, 1962–63.

MODIFIED FROM PANTULU (1986); ORIGINAL TABLE PREPARED BY N.D. LIEN (1983).

in Table 1.2.[176] As shown in Figure 1.3,[177] most of the flow comes from the major left bank tributaries, particularly those that conjoin with the Mekong

176 Pantulu, *supra* note 170, at 708.
177 Gupta and Liew, *supra* note 34, at 262; Gupta, *supra* note 31, at 442.

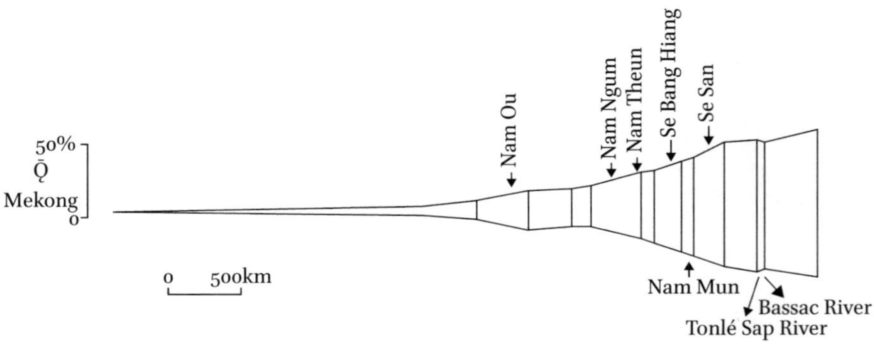

FIGURE 1.3 *Downstream cumulative annual mean flow of the Mekong River. Q = flow. Note the significant volumetric contribution from the left bank tributaries, the Nam Ngum, Nam Theun, and Se Bang Hiang in the Lao* PDR *and the Sesan in Vietnam and Cambodia.*
DATA FROM THE MEKONG RIVER COMMISSION (1994–2004); MODIFIED FROM GUPTA AND LIEW (2007). REPRINTED WITH PERMISSION FROM ELSEVIER.

downstream of Vientiane-Nongkhai.[178] Notable among these major arteries are the right bank Mun-Chi system and the left bank Nam Ngum, Nam Theun, Sesan-Sekong-Srepok system (the "3S Rivers"),[179] and Tonle Sap. Central to this law and policy analysis are the Sesan, Se Bang Fai, Se Bang Hiang, and Nam Ngum rivers, which are considered in the case studies in Chapter 4 of this book.

Some of these rivers flow only in one riparian state, but others, such as the Sesan, which flows through Vietnam and Cambodia, are transboundary tributaries.[180] The Sesan is the second largest tributary of the Mekong River with a drainage area encompassing 76,700 km^2 and is the largest in terms of flow (90,000 million m^3/year; 2,900 m^3/sec); however, these figures include its

178 MRC, *supra* note 45, at 27.

179 Kim Sangha and Tep Bunnarith, "Lessons Learned But Not Learned: Water Governance in the 3S Rivers Region" (Vientiane, Lao PDR, 6–7 July 2006).

180 Philip Hirsch and Kurt Morck Jensen *with* Ben Boer, Naomi Carrard, Stephen FitzGerald and Rosemary Lyster, "National Interests and Transboundary Water Governance in the Mekong" (Australian Mekong Resource Centre (AMRC), University of Sydney in collaboration with Danish International Development Assistance, May 2006). [hereinafter Hirsch and Jensen].

two main tributaries, the Sekong[181] and the Srepok[182] Rivers, which themselves are so large that they are generally treated as separate basins.[183]

The MRC maintains a database of 124 key hydrological stations on the Mekong mainstream and major tributaries. Gauging records at 33 stations on the Mekong, in some cases going back for decades (e.g., at Vientiane back to 1913), are contained in these data with long-term data available for 13 such stations and some discharge measurements from 1924 to the present.[184] The author examined copies of gauging record data on compact discs maintained by the MRC for each year from 1994 through 2004 (11-year record). With these data, the percentage of total flow (discharge) passing through the mouth at each station can be calculated (e.g., 20 percent at station A, 30 percent at station B 1,000 km downstream from station A, etc.). From these data, a cumulative annual mean flow schematic, such as the one prepared by Gupta and Liew (2007) in Figure 1.3, can be plotted where the increase in the width of the diagram reflects the discharge between two stations (Figure 1.3).[185] For example, 80 percent of the cumulative annual mean flow has entered the Mekong when it reaches Stung Treng, Cambodia (Figures 1.1 and 1.3). Hydrologists use this type of diagram to look for explanations of increases in flow, which in the case of the Mekong (as can be gleaned from Figure 1.3), is generally explained by discharge from its left bank tributaries. This increase can be effectively double checked where gauging stations are located directly on tributaries. Most of the flow into the Mekong comes from its left bank tributaries downstream of Vientiane, especially the 3S Rivers which contribute 18 percent at Kratie,

181 Water Environment Partnership in Asia [hereinafter WEPA], *Cambodia, Mekong River tributaries and rainfall*, 2.1.2 <http://www.wepa-db.net/policies/state/cambodia/river2_1_2.htm> (accessed 7 June 2017). The Sekong, which rises in Vietnam, flows through the Lao PDR, conjoins with the Sesan and enters the Mekong at Stung Treng in Cambodia, is one of the largest tributaries of the LMRB with a drainage basin of 28,500 km^2. Three percent of the Sekong watershed lies in Vietnam, 77 percent in the Lao PDR and 20 percent in Cambodia.

182 WEPA, *supra* note 181. Rising in Vietnam and flowing through Cambodia, where 35 km upstream of Stung Treng Province it joins the Sesan, the Srepok River drains an area of 29,450 km^2, 8,200 km^2 of which are located in Vietnam. The Srepok is one of the larger tributaries of the Sesan.

183 Pantulu, *supra* note 170, at 708; WEPA, *supra* note 181.

184 *See, e.g.*, MRC, *Lower Mekong Hydrologic Yearbook* (MRC, 1998) 19–23.; Gupta and Liew, *supra* note 34, at 262.

185 Gupta, *supra* note 31, at 442.

Cambodia, where about 90 percent of the cumulative annual mean flow has entered the river.[186]

The UMRB (China) contributes 16 percent of the flow to the Mekong. The bulk of the river's flow (84 percent) comes from four sources in the LMRB: (1) tributaries draining the mountains of the Lao PDR: (2) the 3S Rivers in the southern Annamite chain; (3) the Mun-Chi rivers in Thailand; and (4) outflow from the Tonle Sap during the dry season.[187] Tributaries provide over half of the flow to the Mekong mainstream.[188] Fifty-five percent of the flow comes from the left bank tributaries in the Lao PDR and the 3S Rivers in the Central Highlands of Vietnam and Cambodia.[189] The flow in these tributaries is closely related to the rainfall pattern associated with the SW monsoon. Runoff from tributaries in the LMRB is 396,000 to 405,000 million m^3 or 80 to 82 percent.[190] Downstream cumulative annual mean flow of the Mekong is shown in Figure 1.3. The importance of the volumetric contribution from tributaries to the river's flow is self-evident, but this contribution is extremely uneven along the mainstream (Figure 1.3).[191] The topographic and geomorphologic characteristics of these streams can produce flashy runoff (Figure 1.4) in the LMRB, resulting in highly variable stream water levels, especially in the rainy season. Mekong tributaries "are extremely seasonal, with steep short-term discharge peaks in the wet season and a prolonged low stage at other times."[192] Significant amounts of water flow from the northern hills of the Lao PDR and northern Annamite chain via the Nam Ngum and Nam Theun systems, and the Central Highlands via the major tributaries of the 3S Rivers (Figure 1.3).[193] Runoff in the Mekong increases significantly with entry from the left bank of the Sesan (Figure 1.3), which along with the Sekong and Srepok, contribute about one fifth of the LMRB flow.

---

186 MRC, *supra* note 45, at 27.

187 Avijit Gupta, "Geology and Landforms of the Mekong Basin" in Ian C. Campbell (ed.), *The Mekong: Biophysical Environment of an International River Basin* (Academic Press, 2009) 29–51, 47–48.

188 Osamu Toda et al., "Evaluation of Tributaries Contribution in the Mekong River Basin During Rainy and Dry Seasons" (Paper presented at the second conference of the Asia Pacific Association of Hydrology and Water Resources, Singapore 5–9 July 2004) 5; Hirsch and Jensen, *supra* note 180, at 15; MRC, "State of the Basin Report 2010" (MRC, April 2010) 20.

189 MRC, *supra* note 188, at 20.

190 Elhance, *supra* note 37 at 195; HORI, *supra* note 2, at 29.

191 Gupta, *supra* note 31, at 439.

192 *Id.*

193 *Id.* at 439–440.

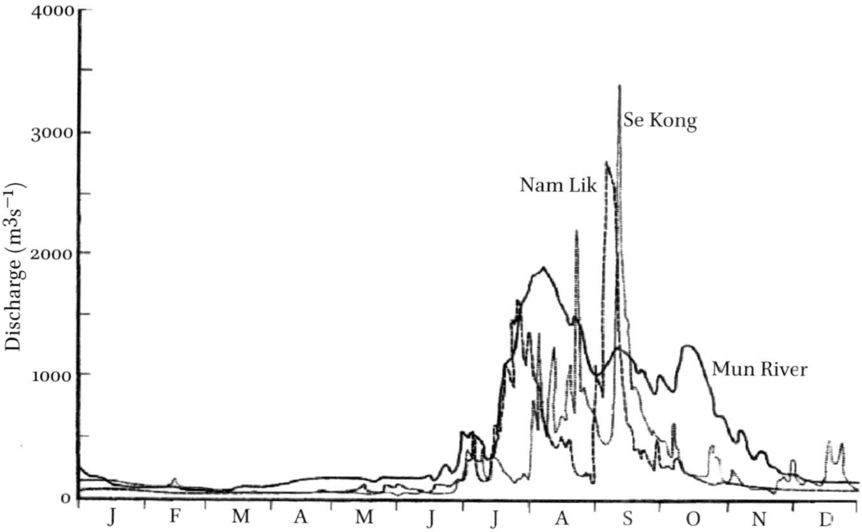

FIGURE 1.4 *Annual hydrographs of selected Mekong tributaries: Nam Lik at Muong Kasi, Lao PDR (basin area 374 km² in northern Lao PDR mountains), Mun River at Ubon, Thailand (104,000 km² on the Korat Plateau) and Sekong at Attopeu (10,500 km² in Annamite Mountains).*
Note the flashy discharge of the left bank Nam Lik and Sekong tributaries. Also note the relatively large volume of discharge on the Nam Lik and Sekong rivers draining the northern Lao PDR and southern Central Highlands. All hydrographs display data from 1997, except the Sekong which displays data from 1993.
SOURCE: *LOWER MEKONG HYDROLOGIC YEARBOOKS*; MODIFIED FROM GUPTA (2007). REPRINTED WITH PERMISSION FROM JOHN WILEY AND SONS, LTD.

Tributary flooding varies across the LMRB.[194] During the SW monsoon, tributaries in Thailand, on the leeward side of the Dong Phya Yen and Petchabun highlands, receive low rainfall.[195] On the flat leeward side in Thailand, tributaries (e.g., Mun River and its tributary the Chi) have relatively small channels and extensive floodplains.[196] Along the windward slopes of the Annamite chain and in the Lao PDR and Cambodia, where higher rainfall occurs, flooding occurs in different magnitudes.[197] On the windward slopes of the highlands of the Lao PDR and Cambodia, streams are geomorphologically "well-defined and

194 Pantulu, *supra* note 170, at 706.
195 *Id.*
196 *Id.*
197 *Id.*

accommodate floods which are fairly uniform from year to year."[198] Mekong tributaries "are profoundly influenced by the seasonal inundation regime."[199]

According to the MRC, there is scant evidence to suggest any systematic changes in the hydrological regime of the Mekong in the last 45 years.[200] Hydrologic data do not indicate any trends of upward or downward shifts in the average magnitude of the flow from 1960 to 2004.[201] Considerable debate continues over whether the effects of upstream reservoir storage in China have altered Mekong dry season hydrology.[202] But what is clear is that in large rivers such as the Mekong, "seasonal flows can be quite variable from year to year."[203] "Although the pattern of the annual hydrograph is fairly predictable, *its magnitude is not*."[204] In any case, Figure 1.3 graphically iterates why there should be increased concern about large-scale reservoirs and water diversions on tributaries and why tributary watercourse law and policy deserve more attention in the Mekong legal regime.

Generally, Mekong tributaries are some of the least developed rivers in the world. Ironically, the majority of large development projects and water diversions in the LMRB have been on these tributaries, particularly in the Lao PDR, Thailand, and Vietnam.[205] The Mekong mainstream has to date not undergone significant development,[206] but this is beginning to change with the continuation of the 28-dam "Lancang cascade"[207] in China and the resurgence of an 11-dam "LMRB cascade."[208]

---

198 *Id.* at 708.

199 V.R. Pantulu, "Fish of the Lower Mekong Basin" in Bryan R. Davies and Keith F. Walker (eds.), *The Ecology of River Systems* (Springer, 1986) 721.

200 MRC, *supra* note 45, at 33.

201 *Id.* These data comport with selected data reviewed by the author in Frits van der Leeden, *Water Resources of the World: Selected Statistics* (Water Information Center, Inc., 1975).

202 Quang M. Nguyen, "Hydrologic Impacts of China's Upper Mekong Dams on Lower Mekong River" (2003); MRC, *supra* note 45, at 33.

203 MRC, *supra* note 45, at 32.

204 *Id.* (emphasis added).

205 Dinar et al., *supra* note 26, at 225.

206 *Id.* at 228.

207 As of 2017, this included seven commissioned dams that are on the Lancang/Upper Mekong, six more that are under construction, and 14 more that are planned. *See* Timo A. Räsänen et al., Observed River Discharge Changes Due to Hydropower Operations in the Upper Mekong Basin (2017) 545 *J. of Hydrology* 28, https://www.thethirdpole.net/2017/02/06/new-study-shows-significant-impact-of-chinese-dams-on-mekong/ (accessed 13 June 2017).

208 Milton Osborne, "River at Risk: The Mekong and the Water Politics of China and Southeast Asia" (Lowy Institute for International Policy Paper 02, Longueville Media, New

Perhaps the most insightful view of the hydrologic link of the Mekong's tributaries to its mainstream and the LMRB to Southeast Asia is that of Evan Gottesman, former Resident Liaison and Deputy Director of the American Bar Association Cambodia Law and Democracy Project. He explains how tributaries conjoining with the Mekong mainstream produced a trade route and immigration pathway which have linked each country of Indochina to the region "geographically, politically, and economically"[209] for centuries. Noteworthy in his analysis is that:

> Cambodia's rivers tell many stories. The Mekong, the longest river in Southeast Asia, originates in the Himalayas and passes through China, Thailand, and Laos before entering Cambodia and heading toward the delta of southern Vietnam. A route for trade and immigration, it links Cambodia to the rest of the region geographically, politically and economically. The Tonle Sap, by contrast, is a purely Cambodian river that flows from a lake of the same name in the center of the country south to Phnom Penh, where, a mere 110 kilometers from its source, it joins the Mekong.
>
> Separately, each river represents a way to look at Cambodia—as a country swept along by regional and global trends or as a unique nation proceeding according to its own traditions. Small and vulnerable, Cambodia has been colonized, overwhelmed by foreign ideologies, torn apart by international conflict, and occupied by other countries. Faced with these incursions, Cambodians have attempted to define an uncontaminated national identity. One version of history thus inspires the other. In riparian terms, the mightier the flow of the international Mekong, the more the Tonle Sap stands as a symbol of Cambodian purity.
>
> Remarkably, this relationship is borne out by reality. For six months every year the Tonle Sap flows south from the lake to its juncture with the Mekong. Then, toward the end of the monsoon rains, the Mekong swells, overwhelming its smaller sister. Over the course of several weeks, the waters of the Tonle Sap begin to churn and then, gradually, reverse course. The river then heads upstream, and the lake expands, flooding the shrubs and trees that skirt its shores and nourishing the fish that teem in its waters.[210]

South Wales, Australia, 2004) 11–13; Milton Osborne, *Under-Threat Mekong River Faces New Dam Challenges* (12 November 2008) The Nation.

209 Gottesman, *supra* note 35, at 12.

210 *Id.* at 12–13.

The Mekong, then, is the "common physical thread" that draws Cambodia, the Lao PDR, Thailand, and Vietnam together into a "single international water resources economy."[211] The Mekong may well have helped to unite the LMRB in the past; however, current practices 20 years on since the adoption of the 1995 Mekong Agreement provide evidence that this mighty river's tributaries seemingly inspire an approach based more on state sovereignty.[212]

1.5.1 *Tributaries in the Context of Holistic Water Resources Management*

Central to the argument that tributaries constitute part of a unitary whole of a river basin and that, accordingly, they should be managed concomitantly with the mainstream, is the concept of holistic water resources management. According to McCaffrey and Weber, sustainable development of water resources requires a holistic approach.[213] A holistic approach "considers all water resources without regard to the existence of international borders."[214] This approach comports with Agenda 21, which provides that "[f]reshwater is a unitary resource" and "[l]ong-term development of global freshwater requires holistic management of resources and a recognition of the interconnectedness of the elements related to freshwater and freshwater quality."[215] "Management of water resources is holistic when it is done on a catchment or drainage basin basis."[216] Because water quantity and quality are inextricably linked in water resources management and because watersheds transcend geopolitical boundaries, integrated water resources management (IWRM) has been proposed by policymakers and scientists to redefine and defragment water management.[217] By implication then, a holistic approach also includes IWRM, which "takes into account not only the ecosystem of which water forms an integral part but also the many different human activities, both existing and proposed, that use and affect freshwater resources."[218]

---

211 MRC, *supra* note 45, at vi.

212 Gottesman, *supra* note 35, at 12.

213 McCaffrey and Weber, *supra* note 6, at 90.

214 Alexandre Kiss and Dinah Shelton, *International Environmental Law* (Transnational Publishers, Inc., 3rd ed., 2004) 458.

215 *Agenda 21*, Division for Sustainable Development, UN Department for Economic and Social Affairs, available at <https://sustainabledevelopment.un.org/content/documents/Agenda21.pdf>, Chapter 18, para. 18.35, p. 287 (accessed 8 June 2017).

216 McCaffrey and Weber, *supra* note 6, at 90.

217 Elli Louka, *Water Law and Policy* (Oxford University Press, 2008) 24.

218 *Id.*

The oft-cited Dublin principles[219] and the Rio Declaration on Environment and Development,[220] as expressed in Chapter 18 of Agenda 21, form the foundation of IWRM. According to Chapter 18 of Agenda 21, entitled "Protection of the Quality and Supply of Freshwater Resources: Application of Integrated Approaches to the Development, Management and Use of Water Resources":

> The widespread scarcity, gradual destruction and aggravated pollution of freshwater resources in many world regions, along with the progressive encroachment of incompatible activities, demand integrated water resources planning and management. Such integration must cover all types of interrelated freshwater bodies, including both surface water and groundwater, and duly consider water quantity and quality aspects.[221]

Agenda 21 proposed program sectors for water resources management which included, *inter alia,* "integrated water resources development and management."[222] IWRM is defined "as a process which promotes the coordinated development and management of water, land, and related resources, in order to maximize the resultant economic and social welfare in an equitable manner without compromising the sustainability of vital ecosystems."[223] McCaffrey and Weber note that IWRM "also has a technical component—the optimal operation of a watershed['s] (or a region's) entire system of water diversion, storage, conveyance, treatment and discharge works."[224] The concept was expanded further in 2002 at the World Summit on Sustainable Development[225] that took place in Johannesburg.[226]

---

219 *The Dublin Statement on Water and Sustainable Development* (Dublin Principles) (International Conference on Water and the Environment, Dublin, Ireland, 26–31 January 1992).

220 *Rio Declaration*, UN Doc. A/CONF.151/5/Rev. 1, 13 June 1992, 31 ILM 874 (UN Conference on Environment and Development (Earth Summit), Rio de Janeiro, Brazil, 3–14 June 1992). [hereinafter Rio Declaration].

221 *Agenda 21, supra* note 215, at Chapter 18, para. 18.3.

222 *Id.,* at para 18.5.

223 Global Water Partnership, "Integrated Water Resources Management" (TAC Paper 4, Global Water Partnership Technical Committee, March 2000).

224 McCaffrey and Weber, *supra* note 6, at 90.

225 World Summit on Sustainable Development (WSSD), Johannesburg, South Africa (26 August–4 September 2002).

226 IWRM was reinforced in para. 26 of the 2002 Johannesburg Plan of Implementation, which refers to the need to develop IWRM and water efficiency plans. According to the Plan of Implementation, countries need to develop integrated water resources and efficiency plans by 2005 including: (1) National and regional strategies, plans and programs

Louka notes in her treatise on water law and policy that:

> According to Agenda 21, IWRM must take place at the level of catchment basin or subbasin and must follow four objectives, including a multisectoral approach to water management, public participation in decision-making, and the strengthening of institutional, legal and financial mechanisms. At the international level, furthermore, states must cooperate in line with existing agreements and arrangements, taking into account the interests of all riparian states.
>
> ....
>
> IWRM is an attempt to address water management based on hydrographic boundaries and organize such management around the river basin district as the coordinating unit regardless [of] the national and administrative borders that may be [in] the way. This means, in practical terms, that upstream states and downstream states have to cooperate to treat a watercourse as the unity nature intended it to be—not piece-by-piece based on artificial boundaries created by states.[227]

Likewise, Kiss and Shelton point out that "[t]he physical unity of an integrated water network should lead naturally to addressing water problems for an entire hydrographic basin."[228] Holistic water resources management or an ecosystem approach, then, is conducted on a watershed or river basin scale.

As early as 1953, Sauser-Hall bemoaned including tributaries in any type of legal ecological unit as a limitation on state sovereignty.[229] Conversely, in the geological context, a river basin is defined as "the entire area drained by a river *and its tributaries*."[230] The modern trend under the UN Watercourses Convention is to treat an international watercourse not just as a single river but as an entire ecosystem, using a definition that has been expanded to include

---

with regard to the integrated basin; (2) a full range of policy instruments; (3) develop programs for mitigating the effects of extreme water-related events; (4) efficient water allocation among competing uses that gives priority to the satisfaction of basic human needs; and (5) facilitation of public-private partnerships. *See* Louka, *supra* note 217, at 27 (citing para. 26 of the 2002 Johannesburg Plan of Implementation).

227 Louka, *supra* note 217, at 26–28 (internal citations omitted).

228 Kiss and Shelton, *supra* note 214, at 455.

229 Georges Sauser-Hall, *L'utilisation Industrielle des Fleuves Internationaux* (Brill, Recueil des Cours, Tome 83, No. 2, 1953) 477–481.

230 Bates and Jackson, *supra* note 22, at 541.

tributaries.[231] As McCaffrey and Weber explain in their Guidebook for Policy and Legislative Development on Conservation and Sustainable Use of Freshwater Resources:

> Increasingly, however, countries are shifting from plans based on either individual watercourses or purely political boundaries to some version of watershed boundaries to demark a planning region's geographic boundaries.
>
> Whether denominated watersheds, catchment basins, or drainage basins, the intent is the same. These devices allow planners to integrate multiple watercourses into a single hydrologically interrelated unit. The impacts of management activities in one part of a basin on resources or uses in another part of the same basin can be more easily identified and addressed when both are part of the same planning activity.[232]

River basins or watersheds, which are functionally equivalent to watercourse systems,[233] "form the most appropriate functional landscape units that can be used to determine the dynamic interactions between land use activities and water resources."[234] Interactions between natural systems and anthropogenic processes "can best be assessed in a watershed context because the water quantity and quality at the outlet of a watershed provides [sic] an excellent indication of how [healthy it is] and what the cumulative impacts are of all … [the watershed] activities."[235] Effective watershed or river basin management requires a holistic approach that inherently includes a watercourse and its tributaries. Whatever the rubric, "[t]hese devices allow planners to integrate multiple watercourses into a single hydrologically interrelated unit."[236]

---

231 1997 UN Convention, *supra* note 18, art. 2(a). As defined in the 1997 UN Convention, an international watercourse is "a system of surface waters and groundwaters constituting by virtue of their physical relationship a unitary whole, and normally flowing into a common terminus."

232 McCaffrey and Weber, *supra* note 6, at 18.

233 McCaffrey, *supra* note 9, at 37. McCaffrey notes that "the concept of the drainage basin is functionally equivalent–at least hydrologically–to that of the watercourse system."

234 University of British Columbia Institute for Resources, Environment and Sustainability, *Watershed Management*, <http://wiki.ubc.ca/Category:Watershed_Management_Certificate> (accessed 8 June 2017).

235 *Id.*

236 McCaffrey and Weber, *supra* note 6, at 18.

1.6 Chapter 1 Conclusion

Rising at over 5,000 m in Qinghai Province, the Mekong, the largest and longest river of Southeast Asia, flows through a geologically diverse basin for over 4,800 km, draining over 795,000 km^2 in six countries. The Mekong landscape is defined, in large measure, by the river and its tributaries which form a river basin *system*. The LMRB is a humid region that contains abundant water resources, half of which emanate from a network of more than 100 major and minor tributaries that are virtually undeveloped. These tributaries are seasonal, driven by the SW monsoon, exhibit flashy discharge, and contribute more than half of the cumulative annual mean flow to the Mekong. China controls the headwaters of the UMRB, which contributes 16 percent of the flow to the Mekong but half of the sediment. The Lao PDR (25 percent) and Thailand (23 percent) share the highest percentage of area in the LMRB, and the Lao PDR contributes the greatest amount of flow (35 percent). The left bank tributaries contribute the lion's share of the cumulative annual mean flow to the Mekong (Figure 1.3). When the Mekong reaches Stung Treng, Cambodia, about 80 percent of the cumulative annual mean flow has entered the river. Particularly noteworthy on the left bank are the Sekong, Srepok, and Sesan rivers, or "3S Rivers," which contribute 18 percent of this flow to the mainstream at Kratie, Cambodia. The Mekong and its tributaries provide a regional geographic, political, and economic link for the LMRB states. Seemingly, these tributaries are emblematic of the national interests of each country.

The importance of tributaries' volumetric flow contributions to the mainstream is self-evident. These contributions establish the significance of tributaries in the Mekong flow regime and reinforce the dangers associated with failing to deal with them adequately in the legal regime and hydropolitical equation in terms of holistic water resources management, which envisions integration of the river *and its tributaries* into a single, hydrologically interrelated unit. Arguably, changes to the flow regimes of tributaries are the greatest emerging challenges at the intersection of international water law and policy in the LMRB because most of the water diversions have taken place on these important river arteries.

CHAPTER 2

Legal and Institutional Framework

2.1 The Mekong Legal Regime: 1957–1995

The evolution of the Mekong legal regime chronicles the United Nations' "first attempt to be directly involved in continuing support for the planning and development of an international river basin."[237] The four states of the LMRB have a history of cooperation spanning six decades.[238] The lauded "Mekong Spirit" has endured the death throes of colonialism, wars of "national liberation,"[239] and even the clashing commitments of ideology evident in the LMRB today.[240]

Commentators generally divide the evolution of the Mekong legal regime into three[241] or four periods.[242] This book subdivides the history of Mekong cooperation into four temporal periods, each reflecting a framework tailored to changes in geopolitical and socioeconomic settings. These periods are as follows: (1) 1957–1975, the Mekong Committee[243] period, the focus of which was the 1957 Mekong Statute[244] and negotiation of the Water Charter;[245] (2)

237 Nakayama, *supra* note 41, at 295.

238 McCaffrey, *supra* note 9, at 285.

239 Khruschev Report, *supra* note 38, at 23.

240 McCaffrey, *supra* note 9, at 285.

241 *E.g.*, Greg Browder and Leonard Ortolano, "The Evolution of an International Water Resources Management Regime in the Mekong River Basin" (2000) 40 *Nat. Resources J.* 499, at 504–530.

242 *E.g.*, Pichyakorn, *supra* note 43, at 171–176; Bennett L. Bearden, "The legal regime of the Mekong River: a look back and some proposals for the way ahead" (2010) 12(6) *Water Policy* 798.

243 Committee for Co-Ordination of Investigations of the Lower Mekong Basin [hereinafter Mekong Committee].

244 Statute for the Committee for Co-Ordination of Investigations of the Lower Mekong Basin, (Cambodia, Laos, Republic of South Viet-Nam and Thailand) Bangkok, Thailand, 17 September 1957 [hereinafter 1957 Mekong Statute]. The draft of the Statute was proposed in 1957 by an *ad hoc* Secretariat established for the Development of Water Resources in the LMRB, UN Economic Commission for Asia and the Far East [hereinafter ECAFE], ECAFE/L.119. Text in UN Legislative Texts and Treaty Provisions Concerning Utilization of International Rivers for Other Purposes than Navigation (ST/LEG/SER.B/12) 267 (1963).

245 Draft Lower Mekong Basin Water Charter (with Commentary), Mekong Committee (Khmer Republic, Laos, Thailand and the Republic of Viet-Nam), ECAFE, Working Paper by the Secretariat, Doc. WRD/MKG/INF/L.427, Bangkok, Thailand, 20 March 1971 [hereinafter Water Charter].

 | DOI 10.1163/9789004362604_004

1975–1978, the Joint Declaration[246] period; (3) 1978–1995, the Interim Mekong Committee[247] period; and (4) 1995-present, the 1995 Mekong Agreement[248] period. In this book, the evolution of the Mekong legal regime is analyzed primarily in the context of its treatment of tributary watercourses.

The LMRB states initially supported a cooperative regime because they expected to benefit from it technologically, economically, and politically. From scientific and engineering perspectives, water resources management has been *the* primary focus in the Mekong regime. Financially, the Mekong regime has been held together by external donors and international assistance.[249]

2.1.1 *1957–1975: The Mekong Committee, the 1957 Mekong Statute and the Water Charter*

The first period in the evolution of Mekong cooperation is referred to herein as the Mekong Committee period. When this period began, Cambodia, the Lao PDR, and Vietnam had emerged from colonialism for only a few years. This period commenced in 1957 amid Cold War politics, under the auspices of UN Economic Commission for Asia and the Far East programs (ECAFE, renamed the Economic and Social Commission for Asia and the Pacific, or ESCAP in 1974), "a regional umbrella organization for the promotion of economic development," and ended in 1975 as a result of the second Indochina War.[250] The focus during most of this period was on formation of the Mekong Committee, establishment of an international legal regime, and procurement of funding.

Established by the governments of Cambodia, Laos, South Vietnam, and Thailand under the 1957 Mekong Statute, the regime's first constitutional document, the Mekong Committee developed as its centerpiece a plan for a Mekong cascade of large dams and reservoirs on the river's mainstream, the impetus of which was strong United States (US) influence. China was not a member of the United Nations in 1957, and accordingly, was excluded, and Myanmar

246 *1975 Joint Declaration of Principles for Utilization of the Waters of the Lower Mekong Basin*, Mekong Committee (Khmer Republic, Laos, Thailand and the Republic of Viet-Nam) Vientiane, Laos, 31 January 1975 [hereinafter 1975 Joint Declaration].

247 Interim Committee for Coordination of Investigations of the Lower Mekong Basin [hereinafter Interim Mekong Committee]; *Declaration Concerning the Interim Committee for Coordination of the Lower Mekong Basin*, Vientiane, Laos, 5 January 1978 [hereinafter 1978 Interim Mekong Committee Declaration].

248 1995 Mekong Agreement, *supra* note 1.

249 McCaffrey, *supra* note 9, at 286.

250 Greg Browder, *Negotiating an International Regime for Water Allocation in the Mekong river basin* (unpublished Ph.D. dissertation, Stanford University, 1998) 33; Browder and Ortolano, *supra* note 241 at 504–509.

(Burma) chose not to participate. The 1957 Mekong Statute focused primarily on economic development in the LMRB and gave scant consideration to environmental and social issues, albeit these were not generally recognized as significant concepts at the time. The mandate was limited to the "the planning aspects of water resource development."[251] According to Article 4 of the 1957 Mekong Statute, the Mekong Committee functions were to "promote, coordinate, supervise and control the planning and investigation of water resources development projects" in the LMRB. The Mekong Committee was empowered to perform four other functions which included the following:

> (a) prepare and submit to participating governments plans for carrying out coordinated research, study and investigation; (b) make requests on behalf of the participating governments for special financial and technical assistance, receive and administer separately such financial and technical assistance, and take title to such property as may be offered under the technical assistance programme of the United Nations, the specialized agencies and friendly governments, or other organizations; (c) draw up and recommend to participating governments criteria for the use of the water of the *main river* for the purpose of water resources development; and (d) employ on behalf of the participating governments, personnel to assist the Committee in the performance of its functions.[252] (emphasis added).

In Chapter 1 of the 1957 Mekong Statute, "Establishment of the Committee" the lower Mekong basin was defined as "namely, the area of the drainage basin of the Mekong river situated in the territory of the participating governments." A key question is whether the undefined term "main river" in Article 4(c) was a harbinger of things to come. Amendments to the 1957 Mekong Statute, revising the Mekong Committee title and expanding its functions (Article 4), were proposed in 1966 but were never ratified.[253] In 1958, The Mekong Committee

---

251 Browder, *supra* note 250, at 39 (1998).

252 1957 Mekong Statute, *supra* note 244, art. 4. (emphasis added).

253 Proposed (unratified) 1966 Amendments to 1957 Mekong Committee Statute, UN ECAFE, Mekong Committee, 29th Session (Special), 12–13 August 1965, revised 29 April 1966, Vientiane, Laos. These Amendments proposed changing the Mekong Committee's title from the Committee for Co-Ordination of Investigations of the Lower Mekong Basin to the "The Committee for Coordination of Comprehensive Development of the Lower Mekong Basin," expanded the Mekong Committee's functions to include "*construction and operation* of water resources development projects ... *and other development projects related thereto*" and allowed the Mekong Committee to "[p]repare and submit to participating

created the Office of the Executive Agent to assist with technical, administrative and funding activities.[254] This office was later referred to as the Mekong Secretariat.[255]

During the 1960s, the Tennessee Valley Authority (TVA) project was emblematic of the ideal program in integrated basin development and was envisioned as a model for the Mekong to "lift the region out of poverty."[256] Accordingly, US President Lyndon Johnson decided to overlay the Mekong with a TVA-type development template.[257] In 1965, in a speech delivered at Johns Hopkins University, Johnson promised $1 billion in development funds for Southeast Asia focusing primarily on a "Mekong project."[258]

Following the publication in 1970 of a 600-page report entitled the Indicative Basin Plan,[259] which was effectively a framework for development projects in the LMRB including a seven-dam cascade on the mainstream, during 1971, the Mekong Committee commenced work on a draft Water Charter. In 1973, the Mekong Committee began formulating a plan for water use and flow when the US offered assistance in drafting rules for water utilization. The Water Charter was, perhaps, the most comprehensive and detailed document ever crafted in the Mekong legal regime as it contained many principles of international watercourse law embodied in the 1966 Helsinki Rules[260] and subsequently included in the UN Watercourses Convention. According

---

Governments plans for carrying out coordinated research, study, investigation [and] *construction and operation*." The Cambodian government did not ratify the amendment. (emphasis added).

254 Browder, *supra* note 250, at 43.

255 *Id.*

256 Browder and Ortolano, *supra* note 241, at 505.

257 *Id.* at 507.

258 *Id.*

259 Mekong Committee, *Report on Indicative Basin Plan: A Proposed Framework for the Development of Water and Related Resources of the Lower Mekong Basin*, E/CN.11/WRD/MKG/L.340, Bangkok, Thailand (un, 1970). The Indicative Basin Plan provided a framework for LMRB development, covered mainstream, tributary and other water resource activities and emphasized a cascade of large reservoirs on the mainstream at Ban Koum, Luang Prabang, Pa Mong, Sambor and Stung Treng.

260 52nd Conference of the International Law Association (hereinafter ILA), Helsinki, Finland, 14–20 August 1966, *Report of the 52nd ILA Conference*. The full text of the Helsinki Rules, with commentary, is contained in International Law Association (ila), *Report of the Fifty-second Conference*, Helsinki, 1966, at 484–532, *reprinted in* ILA, Helsinki Rules on the Uses of the Waters of International Rivers, adopted by the International Law Association at the 52nd Conference held in Helsinki on 20th August, 1966 (1967) [The latter document is referred to hereinafter as the 1966 Helsinki Rules].

to Nakayama, the Water Charter was "a touchstone for the basin countries in terms of their willingness to collaborate, i.e., what they would accept as a duty and what they would wish to enjoy as a right."[261] It contained 74 articles under seven chapters: (1) Definitions, (2) Objectives, (3) Basic Principles, (4) Organization, (5) Financing, (6) Prevention and Settlement of Disputes, and (7) Final Provisions.

Tributaries received detailed coverage in two sections of Chapter 3 of the Water Charter: Section C–Major Tributaries (Articles XXVIII–XXX) and Section D–Minor Tributaries (Articles XXXI–XXXIII). The Water Charter provided that "A Major Tributary is a watercourse which, in its natural or developed state, is recognized by all Basin States as having a determining effect on the regime of the mainstream."[262] In contrast with the mainstream covered by Article X of the Water Charter, the governing concept applied to tributaries remained effectively that of territorial integrity and sovereign equality (in essence hydrosovereignty)[263] with the criterion that major tributaries affecting the Mekong's hydraulic regime would be treated like mainstream waters. Article XXIX of the Water Charter provided that "A tributary recognized by all Basin States as a Major Tributary shall henceforth be considered as an integral part of the Basin development system and shall be governed by the provisions of the present Charter applicable to the mainstream." Article XXX of the Water Charter covered transboundary tributaries and their attendant drainage basins and provided that:

> Unless provided otherwise, a tributary which flows through or separates the territory of two or more Basin States shall be assimilated, by these States, to the mainstream and shall be governed, within their territory, by the provisions of the present Charter applicable thereto.
>
> The drainage basin of such a tributary shall then be considered by the Basin States concerned as a sub-basin and shall be governed by the provisions of the present Charter applicable to the Basin.
>
> In cases where the Basin States concerned so desire, and subject to the concurrence of all Basin States, any such tributary and its basin may be integrated in the Basin development system, in which cases they shall be governed by the provisions of the present Charter applicable to the Basin.

261 Nakayama, *supra* note 41, at 296.

262 Water Charter, *supra* note 245, art. XXVIII.

263 Water Charter, *supra* note 245, pmbl., para. 4.

In the natural progression of holistic water resources management, rules for governance of tributaries should logically comport with the legal framework applicable to the mainstream, since both are critical components of the same river *system*. Ideally then, tributaries should be included in a single, integrated legal framework with the mainstream.[264] Indeed, Article III of the Water Charter captured the spirit of this concept providing that "[t]he water resources of the Basin—in all phases of the hydrologic cycle—constitute a *single natural resource. Each particular utilization of this resource shall be considered in relation to its effect upon the water balance of the Basin.*" (emphasis added).

In the Water Charter, "[a] Minor Tributary is a water course which, in its natural or developed state, has no determining effect on the regime of the mainstream."[265] The LMRB states maintained hydrosovereignty over minor tributaries in the draft Water Charter; however, they could, with the consensus of all basin states, integrate minor tributaries into the mainstream-cum-major tributaries system.[266] Hence, the Water Charter provided for realization of a holistic water resources approach through the concept of the "[b]asin development system"[267] envisioned in Article V.[268]

Because mainstream waters were considered a resource of common interest, no mainstream project, major tributary projects affecting the Mekong's hydraulic regime, or combination thereof, could be implemented under the Water Charter unless all basin states approved it.[269] Implicit, then, in Article XVI of the Water Charter was a veto power over projects on major tributaries. The approach followed in the Water Charter was a balance of hydrology versus sovereignty, wherein tributaries remained under the control of each sovereign state. The underlying view, however, was aimed toward their gradual

264 *See generally,* McCaffrey and Weber, *supra* note 6, at 14–15 and 90, describing the holistic approach to water resources management.

265 Water Charter, *supra* 245, art. XXXI.

266 Water Charter, *supra* note 245, arts. XXXII–XXXIII.

267 Article I(7) of the Water Charter defines" Basin development system" as a "series of projects constructed and operated in such a way as to achieve their hydraulic, electrical and financial integration and to ensure their implementation as components of a single development undertaking."

268 Article V of the Water Charter provides: "To the extent that it is recognized technically feasible, economically justified, socially desirable and consistent with the sovereign rights of the Basin States, individual projects shall be planned and implemented in a manner conducive to the system development of the Basin water resources in the beneficial use of which each Basin State shall be entitled, within its territory, to a reasonable and equitable share."

269 Water Charter, *supra* note 245, art. XVI.

incorporation into the basin development system, anticipating that with sufficient data, most tributaries would be incorporated. This approach fostered a strategy for the LMRB states to manage the hydrological interdependency of the Mekong and its tributaries.

Between 1971 and 1974, a watered-down, revised version of the Water Charter, the Draft Statement of Principles for the Planning and Utilization of the Waters of the Lower Mekong Basin (Draft Statement),[270] was negotiated and completed.[271] The Draft Statement contained 49 articles in five chapters, a significant reduction from the 74 articles in the Water Charter. Conspicuously missing from the revision was the chapter on dispute resolution. The LMRB states later adopted a revised version of the Draft Statement as the 1975 Joint Declaration[272] to replace the 1957 Mekong Statute.[273] In retrospect, the 1975 Joint Declaration proved to be the most progressive agreement addressing tributaries ever signed by the LMRB states, albeit a far cry from the more comprehensive Water Charter.

Other projects and reports during the Mekong Committee era include the 1961 Comprehensive Reconnaissance Report on the Major Tributaries of the Lower Mekong Basin,[274] Preliminary Report on Technical Problems Relating to Flood Control and Water Resources Development on the Mekong—An International River,[275] a report prepared by the US Bureau of Reclamation,[276]

270 Mekong Committee, "Draft Statement of Principles for the Planning and Utilization of the Waters of the Lower Mekong Basin" WRD/MKG/INF/L.466, Bangkok, Thailand (1971) [hereinafter Draft Statement].

271 Mekong Committee, "The 1973 Mekong Committee Report" (1973) 28; Pichyakorn, *supra* note 43, at 172.

272 1975 Joint Declaration, *supra* note 246.

273 Pichyakorn, *supra* note 43, at 172.

274 The Mekong Reconnaissance Team, "Comprehensive Reconnaissance Report on the Major Tributaries of the Lower Mekong Basin" (Mekong Committee, 1961) [hereinafter 1961 Reconnaissance Report].

275 Economic Commission for Asia and the Far East (ECAFE), "Preliminary Report on Technical Problems Relating to Flood Control and Water Resources Development of the Mekong—an International River" (US Bureau of Reclamation, 1952). This report highlighted the immense potential for hydropower, irrigation and navigation in the LMRB.

276 United States Bureau of Reclamation, "Reconnaissance Report—Lower Mekong Basin" (1955). The Bureau's report indicated enormous potential for joint water resource development in the LMRB and advocated damming the Mekong to create large reservoirs for hydropower production, increasing dry season flows, controlling flooding and improving navigation.

and the Ford Foundation study.[277] Also noteworthy during the Mekong Committee era was the establishment of four, state-centric National Mekong Committees (NMCs) in 1957. Charged "with formulating national policy vis-à-vis the Mekong [legal] regime," the NMCs "have remained an important part of the ... regime up to the present time."[278] The Mekong Committee era came to an abrupt end in 1975 when the governments of South Vietnam and Cambodia were overthrown in two wars of national liberation by the North Vietnamese and the Khmer Rouge, respectively. The only major hydropower project completed during the Mekong Committee period, the Nam Ngum dam, ironically, was not on the Mekong mainstream as part of the proposed cascade, but rather on a tributary entirely within the Lao PDR. Today, this dam, which exports power to Thailand, remains controversial.

2.1.2 *1975–1978: The Joint Declaration*

Browder, in his analysis of the LMRB states' negotiations, concluded that the 1975 Joint Declaration was "a milestone in the evolution of the constitutional framework of the Mekong regime for three reasons."[279] First, it expanded the regime's authority from merely planning under the 1957 Mekong Statute to regulation of water resources projects.[280] Second, it became a great source of controversy in the 1990s.[281] Third, the principles embodied in the 1975 Joint Declaration became the focus of negotiations for the 1995 Mekong Agreement.[282]

To be sure, the 1975 Joint Declaration became one of the most important instruments endorsed by the LMRB states because it incorporated in its 35 articles many of the most progressive principles of the law of international watercourses, which were eventually enshrined in the UN Watercourses

277 G. White et al., "Economic and Social Aspects of Lower Mekong Development" (UN Economic Commission for Asia and the Far East, Committee for Coordination of Investigations of the Lower Mekong Basin, 1962). The Ford Foundation study advised the Mekong Committee to focus on smaller tributary projects before developing large reservoirs on the mainstream and cautioned against the full-scale damming of the Mekong without a social and economic analysis. The study warned that water resource infrastructure alone would not produce the hoped-for social and economic benefits to alleviate regional poverty. This sounds like an early forecast of social and environmental issues effectively equivalent to an environmental impact assessment (EIA); Pichyakorn, *supra* note 43, at 172.

278 Browder, *supra* note 250, at 45.

279 *Id.* at 54–55.

280 *Id.* at 55.

281 *Id.*

282 *Id.*

Convention. It provided specific procedures for water use and required approval of all projects by the Mekong Committee prior to implementation. Equitable utilization[283] and pollution prevention[284] were key concepts in the 1975 Joint Declaration. The framework of the 1975 Joint Declaration was subdivided into three sections governing water utilization: (1) Section B–Mainstream, (2) Section C–Tributaries, and (3) Section D–Other water resources, which insightfully included interconnected groundwater.[285] Salient provisions concerning maintenance of water flow and quality[286] and misuse, waste, and pollution prevention[287] were included for the first time in the Mekong legal regime to prevent flow regime alterations and ensure water resources protection. The concept of equality of right was introduced as an allocation method for the Mekong mainstream.[288] In stark contrast to the 1995 Mekong Agreement, in which environmental impact assessments (EIAs) are not a legally binding provision or procedure, the 1975 Joint Declaration required assessments of short- and long-term ecological impacts to other basin states before undertaking mainstream projects.[289] Perhaps most importantly, the Mekong's major tributaries, the arteries which provide significant volumetric contributions to river flow,[290] were treated as mainstream waters.[291] Accordingly, all projects on major tributaries were subject to prior approval by all basin states.[292] In the 1995 Mekong Agreement, planned measures on tributaries may be undertaken with only mere notification to the Joint Committee (JC) of the MRC.[293] As predicted by Stephen McCaffrey, one of the world's foremost authorities on international water resources law and Special Rapporteur for the International Law Commission's (ILC) draft articles for the UN Watercourses Convention, Article 5(A) was destined to cause future problems.[294]

283 1975 Joint Declaration, *supra* note 246, art. V.

284 *Id.*, art. IV.

285 *Id.*, arts. I(2) and XXIII.

286 *Id.*, art. IV(1).

287 *Id.*, art. IV(2).

288 *Id.*, art. XI. Article XI was not interpreted to mean an equal division of the quantity of the water. Compare this with the interpretation of the term equitable share in the US Supreme Court interstate cases of *Kansas v. Colorado*, 206 U.S. 46 (1907) and *New Jersey v. New York*, 283 U.S. 336 (1931).

289 1975 Joint Declaration, *supra* note 246, art. XVII.

290 Toda et al., *supra* note 188, at 3–5; Hirsch and Jensen, *supra* note 180, at 15; MRC, *supra* note 188, at 20.

291 1975 Joint Declaration, *supra* note 246, art. XXI.

292 *Id.*, art. XX.

293 1995 Mekong Agreement, *supra* note 1, art. 5(A). See text of Article 5(A), *supra* note 15.

294 McCaffrey, *supra* note 9, at 287.

As progressive as it was, the 1975 Joint Declaration was never put into practice and unfortunately met an untimely end in that same year, largely due to the geopolitical disruption in Southeast Asia. Cambodia's isolationist foreign policy under the Khmer Rouge communists ended diplomatic relations with its neighbors. Indeed, Thailand was the only state in the LMRB to escape the onslaught of communism. The communist regimes in Cambodia (Khmer Rouge), Laos (Pathet Lao), and Vietnam (communist government under Ho Chi Minh in the North and Viet Cong in the South) chose not to participate in the Mekong regime from 1975–1977, and the continued existence of the Mekong Committee was called into question. Not until 1978 would a new agreement emerge between Laos, Thailand, and Vietnam when they concluded the 1978 Interim Mekong Committee Declaration.

2.1.3 *1978–1995: The Interim Mekong Committee*

A new framework for the Mekong regime based on post-war politics and socio-economic settings was established in 1978 with the formation of the Interim Mekong Committee. The 1978 Interim Mekong Committee Declaration emphatically shifted the regime purpose to promoting "the development of water resources of the lower Mekong basin"[295] from that in the 1957 Mekong Statute of promoting, coordinating, supervising, and controlling the planning and investigation of water resources development projects.[296] The Interim Mekong Committee period was dominated by communist ideology that gave rise to new tensions between Thailand and Vietnam.[297] The Committee had an interim status because Cambodia, under the control of the isolationist Khmer Rouge regime, chose not to participate. The three other LMRB states lacked the spirit of cooperation necessary to proceed with the Mekong cascade project, so construction was postponed indefinitely. Nevertheless, the three states signed the 1978 Interim Mekong Committee Declaration with hopes that Cambodia would someday rejoin the Mekong regime. Prompted by Khmer Rouge border attacks against Vietnam later in 1978, the Vietnamese Army invaded Cambodia, ousting the Khmer Rouge and installing a pro-Vietnamese government.

The Interim Mekong Committee period is considered as a period of interruption of the development of international watercourse law in the LMRB.[298] The principles of the 1975 Joint Declaration were abandoned, no legal

295 1978 Interim Mekong Committee Declaration, *supra* note 247, art. 4.

296 1957 Mekong Statute, *supra* note 244, art. 4.

297 Browder and Ortolano, *supra* note 241, at 509–518.

298 Pichyakorn, *supra* note 43, at 174.

concepts for water use or allocation replaced them, and a "vague outdated framework" similar to the 1957 Mekong Statute was re-adopted.[299] The 1978 Interim Mekong Committee Declaration shifted that Committee's functions to promoting development of water resources in the LMRB, which accordingly, limited Committee actions somewhat.[300] Although the 1978 Interim Mekong Committee Declaration's institutional framework was probably aimed more at maintaining donor confidence and support after withdrawal of Cambodia from the Mekong legal regime, several important environmental programs were initiated by the Interim Mekong Committee, including review of EIA policy and establishment of a water quality monitoring network in 1984.[301]

During the 1980s, Thai and Vietnamese water agencies began pursuing their own, potentially contradictory, national development agendas. In 1987, the Interim Mekong Committee formulated the Revised Indicative Basin Plan. Although the concept of a mainstream cascade endured as a long-term objective, the Interim Mekong Committee period emphatically shifted the focus of development in the LMRB to national projects. Renewed nationalism coupled with China's plan for its own Mekong cascade, a series of eight hydroelectric dams on the Lancang Jiang in Yunnan Province in the UMRB, signaled the end of this period in 1992.

In the post-Cold War era of the 1990s, there was a strong need for redirection in water resources management in the LMRB. Cambodia was seeking readmission[302] into the Mekong regime and the Chinese Langcang cascade was imminent. Seemingly, these events opened up possibilities for new cooperation in the LMRB. The three Interim Mekong Committee states anticipated the reentry of Cambodia into the Mekong legal regime and the Chinese expressed an interest in data sharing with the LMRB riparians. By 1990, the Mekong legal regime's original *raison d'etre* of joint project promotion had become blurred, and the LMRB states began pursuing their respective national water resources agendas.[303] The habits of a profligate past were on an impending collision with ecological and economic limits in the LMRB. The overarching

299 *Id.*

300 1978 Interim Mekong Committee Declaration, *supra* note 247, art. 4. Compare the text of art. 4 of the 1978 Interim Mekong Committee Declaration with art. 4 of the 1957 Mekong Statute which mandated the Mekong Committee to "promote, coordinate, supervise and control the planning and investigation of water resources development projects" in the LMRB.

301 Pichyakorn, *supra* note 43, at 174.

302 Interim Mekong Committee, "Preparing for New Challenges" (1991 Interim Mekong Committee Annual Report, Bangkok, Thailand) 11.

303 Browder, *supra* note 250, at 73.

institutional framework and legal regime treated water as a limitless resource and the myth of abundance was entrenched. Dry season water use and water allocation emerged as the primary strategic natural resources issues for each country. Further, the Mekong legal regime had become a bewilderingly complex administrative galaxy of the procedural rules in the 1957 Mekong Statute, 1975 Joint Declaration, and the 1978 Interim Mekong Committee Declaration. The LMRB states began to understand what developed countries already knew: water flows uphill toward money and one pays for water projects with someone else's money. Consequently, the Mekong regime needed to be restructured.

The pro-western government of Thailand, beginning to realize financial independence and eager to begin national dam projects, was determined to thwart the 1957 Mekong Statute and 1975 Joint Declaration "unanimity rule,"[304] whereby one country could effectively veto another country's water projects.[305] The gravamen of the Thai position was that while waters of the mainstream may be subject to negotiation in light of equitable and reasonable utilization, national sovereignty provides the "highest claim to tributaries of the Mekong that originate in, and flow through Thailand"[306] Thailand claimed the right to allocate water from the mainstream from 12 percent to 16 percent of the total flow, the amount equivalent to the contribution from its tributaries.[307] The Thai Deputy Foreign minister decried the veto rule:

> Under the rules, Thailand's hands are tied by so many requirements that its sovereignty is infringed upon. It's very difficult to have outsiders supervise or even stop the various projects undertaken in our country. This is also impossible from an international legal standpoint. The Northeast, with water being its biggest problem, represents our vital interest .... If the Northeast can not get any share from the abundant resources of the Mekong, where else can we get the water?[308]

---

304 Abigail Makim, "The Changing Face of Mekong Resource Politics in the Post-Cold War Era: Re-negotiating Arrangements for Water Resource Management in the Lower Mekong River Basin (1991–1995)" (Working Paper No. 6, Australian Mekong Resource Centre, University of Sydney, August 2002) 10.

305 Paul Handley and Murray Hiebert, *Hostile Undercurrents: Dispute Deepens Over Use of Mekong River Water* (2 April 1992) Far East Economic Review.

306 Browder, *supra* note 250, at 177. According to Browder, there are no transboundary tributaries in Thailand.

307 Kulachada Chaipipat, *China Still Holds Trump Card on Use of Mekong Resources*, (14 February 1993) The Nation; Weatherbee, *supra* note 46, at 175; Nakayama, *supra* note 41, at 299.

308 Supapohn Kanwerayothin, *The Mekong: More of a Liability Than An Asset?* (21 March 1992) Bangkok Post; Browder, *supra* note 250, at 158–159.

The Thai position was uncompromising. Bangkok fortified its position with the threat of nonparticipation. In reality, all the LMRB states began to place a higher priority on tributary projects than on the vaunted Mekong cascade.

In 1990, Thailand was considering two interbasin water diversion schemes from the Mekong, the Khong-Chi-Mun, and Chao Phraya projects.[309] The former project was slated to begin diverting 90 m^3/sec from the Mekong in 2005 (later increasing to 300 m^3/sec) for an irrigation initiative in Northeast Thailand (Isan) called the "Green Isan" plan and the latter project, with diversions near 500 m^3/sec, was aimed at supplying irrigation water to the Chao Phraya basin and meeting the municipal and industrial water demand for the burgeoning population in Bangkok.[310] Thai officials also wanted China incorporated into the regime because of the Chinese Mekong cascade project's potential impact on Thailand.[311] Vietnam, alarmed that Thai water diversions might potentially affect mainstream flow and harm agriculture via saltwater intrusion into the Mekong delta, advocated a return to the 1957 Mekong Statute and 1975 Joint Declaration rules whereby all proposed projects were subject to unanimous approval.[312] The dry season flow into the Mekong delta is about 2,000 m^3/sec, and Vietnam needs a minimum flow of 1,500 m^3/sec to prevent saltwater intrusion into the delta's rice bowl.[313] This conflict between Thailand and Vietnam almost collapsed the Mekong regime.[314] In 1992, the LMRB states agreed to negotiate a new framework agreement for cooperation on Mekong water resources.[315] Negotiations for the new treaty took place between 1992 and 1995.[316] On 6 October 1992, the UN Development Programme (UNDP) sponsored a consultation to discuss the emerging issues between Thailand and Vietnam and offered to assist as a neutral party.[317] Later that year, 16–17 December, in Kuala

309 Browder, *supra* note 250, at 77–81.

310 *Id.* at 79.

311 *Id.* at 90–91.

312 Browder and Ortolano, *supra* note 241, at 516.

313 NEDECO, "Mekong Delta Master Plan: Thematic Study on Management of Water Resources", "Volume 1: Optimal Use of Water Resources" (MRC, Bangkok, Thailand, October 1993) 2, 42; Browder, *supra* note 250, at 76.

314 Kulachada Chaipipat, *Strong Distrust Delays Cooperation on Mekong* (27 March 1992) The Nation; Browder, *supra* note 250, at 109–110.

315 Browder, *supra* note 250, at 114–115.

316 George E. Radosevich and Douglas C. Olson, "Existing and Emerging Basin Arrangements in Asia: Mekong River Commission Case Study" (Third Workshop on River Basin Institution Development, The World Bank, Washington, DC, 24 June 1999) 1.

317 George E. Radosevich, "Agreement on the Cooperation for the Sustainable Development of the Mekong River Basin: Commentary and History (UNDP, 1995) 2.

Lumpur, Malaysia, the four LMRB states, through their NMCs, set up a Mekong Working Group (MWG) to negotiate and draft a new agreement.[318] This meeting resulted in the drafting of the Kuala Lumpur Communique, a joint framework for future cooperation endorsed by the LMRB states during a meeting with donors on 17–18 December 1992.[319] This Communique set the stage for continued donor funding and increased the LMRB states' commitment to the continuation of the Mekong legal regime.[320]

At the first meeting of the MWG (dubbed MWG-I) in Hanoi on 4–5 February 1993, the four states formally endorsed the Kuala Lumpur Communique, finalized the negotiation protocol, adopted Guidelines for the Working Group on the Future Framework of Mekong Cooperation, approved the selection of Dr. George Radosevich as Senior Advisor (funded by the UNDP), and set a target date of the end of 1993 to complete the MWG tasks.[321] Convened in Bangkok on 4–5 April 1993, the MWG-II meeting was not very productive and the conflict between Thailand and Vietnam re-erupted.[322] The LMRB states agreed to prepare national position papers on the guiding principles of future cooperation and began examining the institutional framework.[323] The LMRB states met next in Vientiane on 28–29 June 1993, where they formulated an outline which they agreed would be transformed into a draft agreement at a Technical Drafting Meeting (TDM).[324] A draft agreement was refined and a new approach to Article 5-Reasonable and Equitable Water Utilization, the most difficult principle to articulate in the negotiations and most contentious provision of the 1995 Mekong Agreement, was proposed at TDM-I held in Bangkok on 17–20 August 1993.[325] The controversy over Article 5 halted progress at the MWG-IV meeting in Phnom Penh on 7–8 October 1993.[326]

At TDM-II in Vientiane on 6–7 January 1994, options were explored concerning reasonable and equitable utilization of the mainstream and tributary waters.[327] Vietnam introduced a new Article 5 which caused a flashpoint and

318 Browder, *supra* note 250, at 120.

319 Radosevich, *supra* note 317, at 2; Browder, *supra* note 250, at 115.

320 Browder, *supra* note 250, at 116.

321 Radosevich, *supra* note 317, at 2; Browder, *supra* note 250, at 149.

322 Radosevich, *supra* note 317, at 2; Browder, *supra* note 250, at 153.

323 Radosevich, *supra* note 317, at 2.

324 Radosevich, *supra* note 317, at 2; Browder, *supra* note 250, at 152.

325 Radosevich, *supra* note 317, at 2; Radosevich and Olson, *supra* note 316, at 13; Browder, *supra* note 250, at 152.

326 Browder, *supra* note 250, at 152.

327 Radosevich, *supra* note 317, at 2.

resulted in the indefinite postponement of the MWG-IV meeting.[328] After meeting in Hanoi in April 1994, the Thai and Vietnamese Prime Ministers reaffirmed the political commitment to move forward.[329] The Indochinese Federation states' (Cambodia, Lao PDR, and Vietnam) MWGs met in Ho Chi Minh City on 7 May 1994, and formulated the joint "May proposal," revising Article 5 for Thailand's consideration.[330] Finally, a draft agreement was initialed by the four MWGs at the MWG-V meeting in Hanoi on 27–29 November 1994, where the Mekong Task Force (transition team) was formed.[331]

On 5 April 1995, the 1995 Mekong Agreement was concluded at a signing ceremony in Chiang Rai, Thailand. The 1995 Mekong Agreement took 21 months to negotiate, superseded the 1957 Mekong Statute and 1975 Joint Declaration, and ushered in the MRC era. China, the regional hegemon, hydro-hegemon[332] and uppermost riparian, and Myanmar chose not to participate in the new legal regime.

2.2 1995–Present: The 1995 Mekong Agreement

The multilateral Mekong Agreement concluded in 1995 and the robust, well-funded joint institution that it erected, the MRC, have been lauded as a "model for the world."[333] According to Radosevich and Olson, the 1995 Mekong Agreement represented "a milestone in international water resource management treaties due to its emphasis on joint development, ecological protection and a dynamic process of water allocation …[,] established the … [MRC,] articulated the principles of cooperation and outlined a set of rules for the reasonable and equitable use of the basin's water resources."[334] The treaty has the concept of sustainable development at its core and was the first transboundary watercourse agreement to adopt this paradigm.[335] However, the mere adoption of a paradigm does not guarantee that it will be implemented. The practice in the Mekong regime illustrates this well as evidenced in Chapters 4 and 6 of this book. Nevertheless, the 1995 Mekong Agreement, the MRC and

328 Browder, *supra* note 250, at 152.

329 *Id.*

330 *Id.* at 152 and 188.

331 Radosevich, *supra* note 317, at 2; Browder, *supra* note 250, at 152.

332 *See generally,* Mark Zeitoun and Jeroen Warner, "Hydro-hegemony—A Framework for Analysis of Trans-boundary Water Conflicts, (2006) 8(5) *Water Policy* 435.

333 MRC, *supra* note 5.

334 Radosevich and Olson, *supra* note 316, at 1.

335 Pichyakorn, *supra* note 3, at 14.

its attendant "Mekong Spirit" seemingly have triumphed to the present day as emblems of what a transboundary watercourse agreement and river basin commission should be. Elhance notes that "[i]n the third World the Mekong basin stands out as a glaring example of the most extensive involvement by the international community in the efforts to develop an international river basin."[336]

The 1995 Mekong Agreement sets forth a general framework for water resources use and development.[337] McCaffrey concisely summarizes the substance of the treaty in his treatise on international watercourses law:

> Areas of cooperation identified in the agreement include irrigation, hydro-power, navigation, flood control, fisheries, timber floating, recreation and tourism. The parties agree to cooperate with regard to these activities "in a manner to optimize the multiple-use and mutual benefits of all riparians and to minimize the harmful effects that might result from natural occurrences and man-made activities." They further agree to protect the environment and ecological balance of the Mekong River Basin, and to "utilize the waters of the Mekong River system in a reasonable and equitable manner." Other substantive provisions relate to such issues as maintenance of flows on the mainstream, state responsibility for damage, freedom of navigation and emergency situations.[338]

These areas of cooperation, then, are the very essence of the 1995 Mekong Agreement, but they also reflect the range of controversial issues facing the LMRB states as these developing countries enter the 21st century, with its attendant environmental challenges at the intersection of international water resources law and policy.

The 1995 Mekong Agreement contains six chapters, the most important of which are Chapters III–Objectives and Principles of Cooperation, IV–Institutional Framework, and V–Addressing Differences and Disputes. Among the international environmental laws and concepts enshrined in the Agreement, the most significant is sustainable development, through which the LMRB states effectively agree to cooperate and make every effort to use their natural resources in a manner in which the environmental conditions and ecological balance of the river are conserved and maintained.[339] Although

336 Elhance, *supra* note 37, at 216.

337 McCaffrey, *supra* note 9, at 285; 1995 Mekong Agreement, *supra* note 1, arts. 1–10.

338 McCaffrey, *supra* note 9, at 286 (internal citations omitted).

339 1995 Mekong Agreement, *supra* note 1, pmbl.

it remains fully binding as a multilateral treaty, the 1995 Mekong Agreement is a broad, basic framework; it is essentially hortatory and "is primarily a co-operative rather than a regulatory agreement."[340] It is, in effect, an agreement to agree. Controversial issues, such as water allocation (Article 5) and flow maintenance (Article 6), were set aside by the LMRB countries during the negotiations of the 1995 Mekong Agreement through the adoption of an "interesting approach."[341] According to McCaffrey, the four countries "captured what they could agree upon in the treaty then included a provision according to which further details regarding certain issues" would be dealt with by procedures.[342] Under the MRC's Water Utilization Program which ran from 2000 to 2008, the following procedures were developed: Procedures for Data and Information Exchange and Sharing,[343] approved in 2001; Procedures for Water Use Monitoring,[344] approved in 2003; Procedures for Notification, Prior Consultation and Agreement,[345] approved in 2003; Procedures for Maintenance of Flows on the Mainstream,[346] approved in 2006; and Procedures for Water Quality,[347] endorsed by the MRC JC in 2006 and finally approved in 2011. These procedures were required to support the Rules that were envisioned, but never promulgated, in accordance with Articles 5, 6, and 26 of the 1995 Mekong Agreement. It is noteworthy that all these "procedures" in the Mekong regime are external to the treaty, and it is questionable whether any of them are legally binding. It is no surprise, then, that commentators such as Lebel et al. (2005), have described the 1995 Mekong Agreement as weak and open to

340 Michael Buxton et al., "Conflict Resolution and Policy Making Mediation in the Mekong River Basin" (September 2006) 41 *Just Policy* 26, 30 [hereinafter Buxton et al. ("Conflict Resolution")].

341 McCaffrey, *supra* note 9, at 285.

342 *Id.*

343 MRC, 1995 Mekong Agreement and Procedural Rules, Procedures for Data and Information Exchange and Sharing, at 20–24 (1 November 2001). Procedure 3, "Principles," states: "The Procedures shall be governed by the following guiding principles:
a. Sovereign equality and territorial integrity;
b. Equitable and reasonable utilization;
c. Respect for rights and legitimate interests;
d. Good faith; and transparency."

344 MRC, 1995 Mekong Agreement and Procedural Rules, Procedures for Water Use Monitoring, at 31–36 (30 November 2003).

345 MRC, 1995 Mekong Agreement and Procedural Rules, Procedures for Notification, Prior Consultation and Agreement, at 38–54 (30 November 2003).

346 MRC, Procedures for Maintenance of Flows on the Mainstream (22 June 2006).

347 MRC, Procedures for Water Quality (endorsed by MRC Joint Committee in 2006 and approved by the MRC Council in 2011).

almost any interpretation, and accordingly, subject to being sidestepped by the LMRB states.[348]

2.2.1 *The Mekong River Commission (MRC)*

In 1995, the MRC replaced the Mekong Committee. The 1995 Mekong Agreement outlines the legal mandate for the MRC in Articles 1–3 and 11. Article 1, defines the areas of cooperation:

> To cooperate in all fields of sustainable development, utilization, management and conservation of the water and related resources of the Mekong River Basin including, but not limited to irrigation, hydro-power, navigation, flood control, fisheries, timber floating, recreation and tourism, in a manner to optimize the multiple-use and mutual benefits of all riparians and to minimize the harmful effects that might result from natural occurrences and man-made activities.

Article 2 relates to projects, programs and planning:

> To promote, support, cooperate and coordinate in the development of the full potential of sustainable benefits to all riparian States and the prevention of wasteful use of Mekong River Basin waters, with emphasis and preference on joint and/or basin-wide development projects and basin programs through the formulation of a basin development plan, that would be used to identify, categorize and prioritize the projects and programs to seek assistance for and to implement at the basin level.

Article 3 mandates environmental protection and ecological balance in the Mekong River Basin: "To protect the environment, natural resources, aquatic life and conditions, and ecological balance of the Mekong River Basin from pollution or other harmful effects resulting from any development plans and uses of water and related resources in the Basin." Finally, Article 11 identifies the MRC as the joint basin institution through which cooperation is to be realized:

> The institutional framework for cooperation in the Mekong River Basin under this Agreement shall be called the Mekong River Commission [MRC] and shall, for the purpose of the exercise of its functions, enjoy

348 *See* Louis Lebel et al., "The Politics of Scale, Position, and Place in the Governance of Water Resources in the Mekong Region" (2005) 10(2) *Ecology and Society*, article 18, available at <http://www.ecologyandsociety.org/vol10/iss2/art18/> (accessed 12 June 2017) [hereinafter Lebel et al. (2005)].

the status of an international body, including entering agreements and obligations with the donor or international community.

Analysis of these four articles leads to the characterization of the MRC as an organization erected for sustainable river basin management rather than as an agency charged with promoting megaprojects.[349] Interpreted in its broadest terms in the context of the 1995 Mekong Agreement, the role of the MRC is to promote and coordinate sustainable development of water and related resources in the basin. In 2005, the MRC defined Integrated Water Resources Management (IWRM) as the approach for implementing sustainable development.[350] Clearly, through Articles 1–3 of the 1995 Mekong Agreement, the MRC is mandated to engage in a leadership role in transboundary and basin-wide water resources management and development in the LMRB. Although the MRC is supposed to provide guidance for the sustainable development of the water resources in the LMRB, perhaps the most striking feature of the Mekong legal regime is the MRC's lack of a mandate to act solely on its own without approval of the member states.

The MRC consists of three permanent bodies: the Council, the JC, and the Secretariat.[351] The 1995 Mekong Agreement entrusts to the Council regulatory authority, and policy- and decision-making power with a view to implementing the Mekong legal regime.[352] The Council, the highest-ranking body of the MRC, has one member from each MRC member state at the Cabinet or Ministerial level.[353] It meets annually[354] and develops policies, frames issues, and provides overall guidance of the MRC.[355] The JC is composed of one member from each member state at the Department head level.[356] The JC meets biannually,[357] implements the Council's decisions, and also has some rather

349 The MRC envisions itself as "[a] world class, financially secure, international river basin organization serving the Mekong countries to achieve the basin vision,' which is "[a]n economically prosperous, socially just and environmentally sound Mekong River Basin." MRC 2004 Annual Report 2. The overarching mantra of the 1995 Mekong Agreement and the MRC is "meeting the needs, keeping the balance." MRC, Comments by Mr. Do Manh Hung, Officer-in-Charge, MRC Secretariat (Press Statement regarding the press briefing held on 13 November in Bangkok, Thailand by (TERRA), 15 November 2007).

350 MRC, "Strategic Plan: 2006–2010" (December 2006), 21–23.

351 1995 Mekong Agreement, *supra* note 1, art. 12; Pichyakorn, *supra* note 43, at 175.

352 1995 Mekong Agreement, *supra* note 1, art. 18.

353 *Id.*, art. 15.

354 *Id.*, art. 17.

355 *Id.*, art. 18.

356 *Id.*, art. 21.

357 *Id.*, art. 23.

limited arbitral and judicial functions.[358] The Secretariat is supervised by the JC and headed by a Chief Executive Officer (CEO),[359] who is a foreigner from one of the external donor countries, a feature of the MRC that is unique but augers well for neutrality.[360] It renders technical and administrative support to the JC and Council, formulates an annual plan and report and maintains and updates a technical information database.[361] The Secretariat's divisions are headed by a member state director, with the exception of Finance and Administration. The Secretariat works with the NMCs, which coordinate MRC programs at the national level and provide links between the MRC Secretariat and the national ministries and line agencies in each LMRB state. The NMCs represent their states on the JC and guard their turf zealously.[362] China and Myanmar are not members of the MRC but hold observer status.

The original MRC Mekong Program included several components.[363] The MRC initially approved three programs: Core Programs, Support Programs, and Sector Programs. The Core Programs include the Water Utilization Program, Basin Development Plan, and the Environment Program. The Support Program focused on capacity building for the staff of the MRC knowledge management and national administrative authorities relating to the implementation of the Mekong program framework. The Sector Programs consisted of the Hydropower, Drought Management, Flood Management and Mitigation, Fisheries, Agriculture, Irrigation, and Forestry (including the Watershed Management Project), Navigation, and Tourism program elements.

Under its 2016–2020 Basin Development Strategy and Strategic Plan, the MRC Council began the implementation of a "Roadmap for Decentralisation," which will reorganize the duties of the Secretariat from the original program-based structure, and replace it with a corporate-style structure, incorporating many administrative objectives of the former programs into core functions of the Secretariat divisions.[364] Meanwhile, the responsibility to finance and administer many of the specific functions of the former programs

358 *Id.*, art. 24(F).

359 *Id.*, art. 31.

360 McCaffrey, *supra* note 9, at 286.

361 1995 Mekong Agreement, *supra* note 1, arts. 28 and 30.

362 John Dore and Kate Lazarus, "De-marginalizing the Mekong River Commission" in Francois Molle et al. (eds.), *Contested Waterscapes in the Mekong Region: Hydropower, Livelihoods and Governance* (Earthscan, 2009) 357, 359.

363 MRC, The Mekong Programme, *supra* note 130, available at <">http://archive.iwlearn.net/www.MRCmekong.org/www.MRCmekong.org/mekong_program_ceo.htm#integrated_water> (accessed 12 June 2017).

364 MRC, "Strategic Plan 2016–2020" (March 2016) 9 [hereinafter 2016 Strategic Plan].

will be shifted to individual member countries.[365] The organizational change was stimulated by the perceived need to enhance efficiency and effectiveness through a "leaner river basin organization."[366] At the same time, however, the 2016–2020 Basin Development Strategy (BDS) continues to place key importance on the IWRM concept; indeed, the title of the document begins with "Integrated Water Resources Management-based," indicating the importance of the concept in developing the BDS.[367]

Of central importance in the responsible development of Mekong tributaries are the five sets of "MRC procedures" (formerly encapsulated as part of the Water Utilization Program), which the MRC and member countries found were not being implemented effectively under the prevailing program structure.[368] These procedures, which will be henceforth implemented by the member countries, are (1) Procedures for Data and Information Exchange and Sharing; (2) Procedures for Water Use and Monitoring; (3) Procedures for Prior Consultation and Agreement (PNPCA); (4) Procedures for Maintenance of Flows on the Mainstream; and (5) Procedures for Water Quality.[369] As expressed in the BDS, the implementation of the MRC procedures was an important factor in the decision to decentralize and reorganize the program structure. Specifically, the MRC found that, concerning the Procedures for Data and Information Exchange and Procedures for Water Use Monitoring, greater clarity was needed regarding data used when executing MRC-sanctioned activities.[370] Additionally, BDS advanced the need for the advent of a "practical water quality management tool" to make the Procedures for Water Quality simpler and more effective.[371] Finally, the MRC expressed that deficiencies in the PNPCA were made evident by the tensions that occurred during negotiation processes for the Xayaburi and Don Sahong dam projects on the Mekong mainstream in the Lao PDR.[372] It will remain to be seen whether the decentralization strategy advanced by the MRC in the BDS and 2016 Strategic Plan will result in better governance, and whether the member countries will be able to effectively implement the MRC Procedures.

365 *Id.* at 10.

366 *Id.* at xiv.

367 *See* MRC Council, "Integrated Water Resources Management-based Basin Development Strategy 2016–2020 For the Lower Mekong Basin" (2016) [hereinafter BDS 2016–20].

368 *See generally, id. See also,* 2016 Strategic Plan, *supra* note 364.

369 *See* MRC, *supra* notes 343, 344, 345, 346, 347.

370 BDS 2016–20, *supra* note 367, at 11.

371 *Id.*

372 *Id.*

A key question to consider is whether the MRC has lived up to its reputation for imagination, resourcefulness, and spirit as it meets the challenges of holistic water resources management and basin-wide governance in the LMRB. Due largely to unbridled development and planned water diversions on the Mekong and its tributaries[373] and a "galvanized … environmentalism,"[374] the MRC is suffering a crisis of relevancy and legitimacy and may be at crossroads. Some commentators argue that the MRC could play an important role in the decision- and policy-making process in the LMRB, but often the member states and NMCs frequently by-pass it and accordingly, it is underutilized.[375] Case studies on hydropower development on the Sesan River and the proposed water diversions in the Thai water grid, discussed in Chapter 4 of this book, and the Agreement on Commercial Navigation Along the Upper Reach of the Mekong River[376] evidence such sidestepping and marginalization of the MRC. The Chinese "Mekong cascade"[377] effectively lops off the headwaters of the Mekong, disdaining the multilateral approach of the 1995 Mekong Agreement and the MRC, especially because China prefers to deal with water resources management unilaterally or bilaterally. This is not surprising given the Chinese contempt for and vote against the UN Watercourses Convention. The MRC had little involvement in any of these dam, navigation, and water diversion megaprojects beforehand, during the planning phase, or in the aftermath. Worse, perhaps, is the observation by Hirsch that "there is no real consistency between the [1995] Mekong Agreement and the national water policies legislation of the states that are signatory to it."[378] Hirsch observes the problems that can be caused by emphasis on national interests:

---

373 Fred Pearce, *When the Rivers Run Dry: Water—The Defining Crisis of the Twenty-First Century* (Beacon Press, 2006) 101.

374 Philip Hirsch, *supra* note 8, at 312.

375 Dore and Lazarus, *supra* note 362, at 358–363.

376 The Agreement on Commercial Navigation Along the Upper Reach of the Mekong River is the multilateral compact signed by China, Myanmar, the Lao PDR and Thailand which requires, *inter alia*, removal of navigational obstacles. *See generally* Kulachada Chaipipat, *Mekong River Pact to Be Signed Today* (19 April 2000) The Nation, available at <http://www.burmalibrary.org/TinKyi/archives/2000-04/msg00015.html>. (reporting on the historical pact to be signed by officials from China, Burma, Thailand, and Laos) (accessed 12 June 2017); Dore and Lazarus, *supra* note 362, at 361–362.

377 *See generally* Dore and Lazarus, *supra* note 362, at 362–363. (China's "Mekong cascade" is a series of eight hydroelectric dams on the Lancang Jiang (Mekong mainstream in China) in Yunnan Province).

378 Philip Hirsch, "The Mekong River Commission and the Question of National Interest(s)" (2006) 12(1) *Watershed* 20, 22.

> In a large transboundary river basin such as the Mekong, the level of complexity is even greater than in smaller basins within individual countries. This is both because of the very scale of the basin and because of the political and sovereignty issues involved in international cooperation and shared use of the basin. Ironically, the institutional framework established to deal with transboundary management can lead to simplification as water sharing arrangements are defined in terms of one country's national interest versus that of another, as if national interest can be ascertained and represented in an uncomplicated way. While there are clearly important sovereignty and border issues to address, many of the problems of competing uses and values occur at a societal level within countries, and these issues can be neglected when the sovereign country representation in institutional arrangements is limited to a particular set of interests. This situation can be further exacerbated when those interests are primarily "developmental," in the sense of seeing river basin cooperation as essentially a task of mobilizing resources that allow for construction of irrigation and hydropower infrastructure to harness "underutilized" water—without taking sufficient account of the implications of such use for those whose fishing and farming livelihoods depend on the river.[379]

Further, Hirsch questions the MRC's role in enforcing the 1995 Mekong Agreement:

> The regulatory role of the MRC is debatable. According to the MRC itself, the [1995] Mekong Agreement is not a regulatory instrument and the MRC is not a regulatory authority. On the other hand, its core programmes are intended to build a knowledge base in support of rules governing water sharing, transboundary impact assessment, and the prioritizing of development plans. Some people in the MRC do acknowledge that this amounts to "some degree of regulation."[380]

He urges the MRC to move from hoary nationalism to a forward-thinking approach:

> The challenge for the MRC is to move beyond its current "woolly thinking" on regulation and governance. As a transboundary agency, it must

379 *Id.* at 20.

380 *Id.* at 23–24.

> transcend narrow national agendas and embrace a more representative range of national interests, not least the interest of basin communities and stakeholders.[381]

Hirsch posits that the LMRB states, through the MRC, should garner the political will to move the Mekong legal regime from soft law to hard law, in order to strengthen their commitment to IWRM and not divorce themselves from basin flashpoints such as hydropower development on the Sesan River.[382]

Dore suggests that some state actors of the LMRB states prefer a paper tiger MRC that concentrates on project opportunities and fundraising, while leaving development to the NMCs.[383] According to Dore and Lazarus, the MRC "has a contested governance mandate—embodied in the 1995 Mekong ... Agreement—for the mainstream, tributaries and the lands of the basin within the territories of the Lower Mekong countries."[384] They maintain that "[o]n ... tributaries, disputes exist, resulting from interventions to natural flow regimes and overt or default management decisions" and that "a more deliberate water politics" is needed.[385] National agendas have thus far prevailed over "deliberation" which Dore and Lazarus define as follows:

> Deliberation is debate and discussion aimed at producing reasonable well-informed opinions in which participants are willing to revise preferences in light of discussion, new information, and claims made by fellow participants. Although consensus need not be the ultimate aim of deliberation, and participants are expected to pursue their interests, an overarching interest in the legitimacy of outcomes (understood as justification to all affected) ideally characterizes deliberation.[386]

Dore and Lazarus envision "[a] new water governance paradigm" in which contested projects are taken off negotiating tables and placed on deliberation schedules with the MRC and all stakeholders fully and wholly engaged.[387]

---

381 *Id.* at 24.

382 *Id.* at 24–25.

383 John Dore, "The Governance of Increasing Mekong Regionalism" in M. Kaosa-ard and J. Dore (eds.), *Social Challenges for the Mekong Region* (White Lotus, 2003) 425.

384 Dore and Lazarus, *supra* note 362, at 358 (citations omitted).

385 *Id.* at 357.

386 *Id.* at 358 (citing S. Chambers, "Deliberative Democratic Theory" (2003) 6 *Annual Review of Political Science* 307, 309).

387 Dore and Lazarus, *supra* note 362, at 357.

During its 20-year history, the MRC Secretariat has had four CEOs, all of whom had interests as disparate as the state-centric NMCs. According to Dore and Lazarus, this period has been "punctuated with caretaker leadership."[388] From 1995 to 1998, Yasunobu Matoba, the first CEO of the MRC Secretariat, presided over a state-centric basin institution which lacked command performance and followed the money of "often independently operating donor projects."[389] His successor, Joern Kristensen (2000–2003), shifted the emphasis of MRC management to that of a "knowledge broker ... with a strong commitment to improving the livelihoods of the people of the basin."[390] Olivier Cogels (2004–2007), the third CEO, envisioned two lofty goals for the MRC Secretariat: (1) to build a relationship with China; and (2) to become an investment promoter or facilitator.[391] Dore and Lazarus observed that "[b]oth of these moves brought him into conflict with the knowledge-brokering role as the promoter/facilitator was uninterested in any bad news about possible negative impacts of ... tributary development."[392] A review team for the MRC Secretariat stated the "MRC S[ecretariat] is starting to become known among civil society organizations, scientific organizations, etc., as an institution that will not release information that may illustrate negative environmental and social consequences of development projects. This is a threat to the credibility of the organization."[393]

Dore and Lazarus posited that the fourth CEO, Jeremy Bird, who assumed the reins in 2008, came to the position with a "solid grasp" of Mekong hydropolitics, and had an opportunity to open new vistas to demarginalize the MRC.[394] In his introductory speech toasting cooperation in the LMRB, Bird proposed four central themes for the MRC Secretariat's future efforts all beginning with the letter R (as in "River"): Regionalism and riparianism, relevance, responsibilities, and risk reduction.[395] To these four "Rs", Dore and Lazarus emphatically added "examining rewards (winners and losers)" and "respecting rights."[396] In

388 *Id.* at 366.

389 *Id.*

390 *Id.* at 366–367.

391 *Id.* at 367.

392 *Id.*

393 N. Hawkesworth et al., "Independent Organisational Financial and Institutional Review of the MRC Secretariat and the National Mekong Committees" (MRC, Vientiane, Lao PDR, 2007) 20.

394 Dore and Lazarus, *supra* note 362, at 373.

395 Jeremy Bird, "Introductory Speech" (Speech delivered as new MRC Secretariat CEO, Vientiane, Lao PDR, 7 April 2008).

396 Dore and Lazarus, *supra* note 362, at 376–377.

the author's judgment, other "Rs" that should have been considered for interjection into CEO Bird's central themes are a "reformed" mandate for the MRC and "refusion" of the lifeblood of the Mekong, its tributaries' waters, into the legal regime.

After Jeremy Bird, Hans Guttman served as MRC Secretariat CEO from 2011–2015, and served as the head of the MRC Environment Programme from 2001–2007.[397] During Guttmans's tenure, the MRC (and occasionally Guttman himself) received multiple criticisms regarding its prior consultation and negotiation process as toothless and unable to curtail profligate national interests after controversial hydropower projects along the Mekong mainstream in the Lao PDR ultimately went forward, leaving many stakeholders feeling discontent and overlooked.[398]

On 18 January 2016, MRC Secretariat CEO Pham Tuan Phan began his three-year term. The appointment of Phan has been largely touted as falling in line with the restructuring and decentralization of the MRC, and he will take a leading role in implementing the reforms in the 2016 Strategic Plan going forward.[399] Phan's appointment is particularly significant because it marked the first time that the MRC appointed a CEO from one of its member nations (Phan hails from Viet Nam), a move that is further emblematic of the recent "riparianization" of the MRC.[400]

There is no single entity that has plenary jurisdiction over basin-wide governance of the whole LMRB.[401] Although heralded as a "step towards overcoming past antagonisms" in the LMRB, "the establishment of the MRC obscured some very real problems."[402] Noted Mekong commentator and Lowy Institute Fellow Milton Osborne observed that:

---

397 MRC, *MRC appoints new Chief Executive Officer* (Press Release, November 14, 2011) http://www.MRCmekong.org/news-and-events/news/MRC-appoints-new-chief-executive-officer/ (accessed 19 September 2017).

398 These criticisms may be unfounded, at least in part, because they fail to recognize that the MRC is not a regulatory body by design. *See, e.g.,* Tom Fawthrop, *Mekong Summit Struggles to Halt Devastating Dams* (9 April 2014) The Diplomat <http://thediplomat.com/regions/east-asia>; Tom Fawthrop, "Death by strangulation? Hydropower threatens to kill the mighty Mekong" (27 March 2015) The Ecologist <http://www.theecologist.org/News/news_analysis/2806736/death_by_strangulation_hydropower_threatens_to_kill_the_mighty_mekong.html> (accessed 19 September 2017).

399 MRC, *First Riparian Chief Executive Officer Assumes His Office Today* (Press Release, 18 January 2016) < http://www.MRCmekong.org/news-and-events/news/first-riparian-chief-executive-officer-assumes-his-office/> (accessed 18 September 2017).

400 *Id.*

401 Osborne (2004), *supra* note 208, at 5.

402 *Id.* at 7.

> Since its establishment, the MRC has been an almost continuous target of criticism from advocacy NGOs [non-governmental organizations) and academic observers for its supposed failure to take an active role in addressing the range of environmental challenges facing the Mekong River. Such criticism is misguided since it fails to recognize *the MRC is a "creature" of the governments that are members of the Commission. While its critics might want it to be otherwise, the MRC has no mandate to act on its own in any fashion that has not been approved by the member countries.*[403]

Lebel and Garden regard the marginalization of tributaries in the Mekong legal regime, despite the fact that some are transboundary watercourses, as "a good example of politics of place and scale" which contributes to the "complex, vague and inconsistent" wording in the current treaty.[404] They note that dams are being built in China, the Lao PDR, and the highlands of Vietnam "with little reference to multilateral frameworks like the MRC."[405] Accordingly, they posit that "the 1995 Mekong Agreement can be read as a dead letter,"[406] and refer to the MRC as "an amputated river basin organization with its tributaries pruned and its headwaters lopped."[407]

Unless the MRC's legal mandate is reformed, Osborne's and Lebel and Garden's observations might well serve as a prescient outlook. A case in point is the current environmentalism and opposition to the resurrection of the plans for the controversial Mekong mainstream cascade in the LMRB originally proposed in the 1950s during the Mekong Committee period.[408] Since mid-2006, the four LMRB states have been investigating proposals for 11 dams on the Mekong mainstream located at Pak Beng, Luang Prabang, Xayaburi, Pak Lay, and Sanakham in the northern Lao PDR; Pak Chom and Ban Koum on the Thai-Lao PDR border; Lat Sua and Don Sahong in the southern Lao PDR; and Stung Treng and Sambor in Cambodia.[409] The Xayaburi Dam was 70 percent complete as of May 2017, and construction has recently begun on the Don Sahong project.[410] The Xayaburi project became the first mainstream dam to undergo the PNPCA in September 2010, but the Lao PDR government ultimately decided to begin construction without addressing the recommendations of the MRC's

403 *Id.* at 9 (emphasis added).

404 Lebel and Garden, *supra* note 20, at 209–210.

405 *Id.* at 212.

406 *Id.* at 211.

407 *Id.* at 216.

408 IRN, *Mekong Mainstream Dams* (2016), available at: <http://internationalrivers.org/en/southeast-asia/mekong-mainstream-dams> (accessed 13 June 2017).

409 *Id.*

410 *Id.*

Strategic Environmental assessment.[411] In addition to similar issues with fully following the PNPCA process prior to the construction of the Don Sahong Dam, no transboundary assessment was conducted, despite the fact that the project is located 2 km from the Lao-Cambodia border.[412] When prompted with some of these criticisms, former MRC Secretariat CEO Hans Guttman clarified that "the MRC is not a regulatory body. We can only facilitate dialogue between member states, we cannot enforce anything ... prior consultation is not a process to seek approval for a proposed project."[413] Thus, the Lao PDR government did not technically violate the process by refusing to consider whether the project should be discontinued and instead simply offering to mitigate some of the adverse impacts. This may represent a weakness in the previous administration of the PNPCA. The restructured implementation of the PNPCA under the decentralization strategy and leadership of CEO Phan will certainly be tested, especially as the Prior Consultation subprocess for the Pak Beng project is currently underway.

Because the MRC is no longer the only institutional mechanism in the region to promote economic development, additional partnerships have emerged with scopes overlapping with the MRC, such as the one with the Greater Mekong Subregion, a program initiated by the Asian Development Bank that includes all six Mekong basin states. The Bank is a donor to the MRC and has observer status in it. The Greater Mekong Subregion (GMS) not only deals with water sector issues but also with broader issues such as navigation, in which the Chinese have a vital interest. For instance, initiatives of the GMS program's Energy Roadmap may potentially run counter to interests of certain stakeholders, as it highlights the significant hydropower development potential in the Lao PDR and China's Yunnan Province as an important part of sustainably developing the GMS countries' energy portfolios.[414] At the same time, the report also emphasizes the importance of analyzing social and environmental costs to transboundary nations and downstream stakeholders, which would

411 *Id.* Namely, the SEA recommended a deferment on decisions of whether to proceed with hydropower projects on the Mekong mainstream for ten years until additional studies could be conducted, which would allow potentially adverse environmental and hydrological impacts to be better understood. There were also numerous concerns from downstream countries that the Lao PDR did not address in updating the design of the Xayaburi Dam.

412 *Id.*

413 Fawthrop (2015), *supra* note 398.

414 Asian Development Bank, "Building a Sustainable Energy Future: The Greater Mekong Subregion" (GMS Report, 2009) 9. *See also,* Asian Development Bank, *GMS Sector Activities* <https://www.adb.org/countries/gms/sector-activities> (accessed 14 June 2017).

seem to place the Asian Development Bank in indirect competition with the MRC as the multilateral forum for evaluating these costs and benefits.[415]

The question for the MRC in light of these additional regional arrangements is whether it can remain relevant in transboundary water governance given its limited mandate to act only with approval from member states' NMCs. According to Backer, this side-by-side existence of the Greater Mekong Subregion and the MRC "has [already] at times been difficult both for riparians and the involved external donors."[416]

2.2.2 *National Interests and Water Law*

When considering tributary watercourse law, the national interests of the LMRB states should be weighed relative to the MRC framework, NMCs, and domestic water law and policy. Analysis of the water legislation and water agencies' frameworks reveals a conspicuous lack of any meaningful incorporation of the 1995 Mekong Agreement or reference to the role of the MRC in the national water law and policy of the LMRB states. This state of affairs is all too telling, especially because the role of the NMCs is to coordinate MRC programs at the national level and provide a primary point of contact between the Secretariat and domestic agencies. Indeed, 20 years on from the signing of the 1995 Mekong Agreement, most domestic water laws and policies are not well developed. Simply stated, specific provisions to implement the current Mekong legal regime do not exist in most jurisdictions.

There was no specific water resources legislation in Thailand at the initiation of this research but the Water Resources Bill (draft Water Act), ten years in the making, had been submitted to the cabinet for approval. Thailand lacks a comprehensive national policy or strategy for holistic water resources management that incorporates a commitment to the substantive and procedural elements of sustainable water development.[417]

The draft Water Act provides a mandate for river basin committees to approve basin development plans for interbasin transfers to overcome "natural allocation." Separate laws on groundwater, irrigation, and water quality evidence a

415 Asian Development Bank, *supra* note 414, at 21.

416 Ellen Bruzelius Backer, "The Mekong River Commission: Does It Work, and How Does the Mekong Basin's Geography Influence Its Effectiveness?" (2007) 26(4) *Journal of Current Southeast Asian Affairs* 31.

417 The substantive and procedural elements of sustainable development and sustainable water development are provided in Section 2.2.6 of this book. *See generally,* Patricia W. Birnie and Alan E. Boyle, *International Law and the Environment* (Oxford University Press, 2nd ed., 2003) 85–95; Pichyakorn, *supra* note 3, at 11.

need for conjunctive water resources management, generally, and the lack of consideration of hydrological interconnectivity, specifically. Although a party to the 1995 Mekong Agreement, Thailand has no specific law or policy providing for cooperation in managing shared water resources. If adopted, the draft Water Act creates the National Water Resources Committee and river basin committees and mandates water management, water allocation, dispute resolution, and stakeholder participation.

Thailand's national interests in the "Thai water grid" focus on large-scale water diversions and interbasin transfers to the dry Northeast area (Isan) of the country for irrigation initiatives and to Bangkok for potable water supplies and to a more limited extent on hydropower.[418] As a downstream riparian to China, Thailand is also concerned about the "Mekong cascade" on the Lancang Jiang. Irrigation projects are overseen by the Royal Irrigation Department. Basin and sub-basin planning is under the jurisdiction of the Department of Water Resources within the Ministry of Natural Resources and Environment. These agencies lack a coordinated and cooperative policy, especially in the heavily criticized Office of the National Water Resources Committee.[419] Thailand is in dire need of an integrated national water resources strategy. In 2013, there was a draft law in Thailand for IWRM, which sought to integrate the functions of the myriad of government agencies that deal with water. This draft law was developed in a grassroots capacity with International Union for Conservation of Nature (IUCN) and several NGOs. Although the principles of IWRM are not institutionally provided for via comprehensive legislation or policy in Thailand, the country has been lauded for the ability to implement some facets of IWRM in practice. Gao notes that the underlying structure provided by its 25 river basin committees, as well as a relatively enthusiastic and organized environmental movement, bestows Thailand with a capability for comparatively strong stakeholder participation and citizen engagement.[420] However, Gao adds the caveat that these conditions do not necessarily extend to the Thailand's behavior at a transboundary level, as the national government and many of the stakeholders in its environmental movement have no qualms with purchasing electricity produced by controversial hydropower facilities outside of the nation's borders.[421]

---

418 Dinar et al., *supra* note 26, at 234.

419 Hirsch and Jensen, *supra* note 180, at 35.

420 Qi Gao, *A Procedural Framework for Transboundary Water Management in the Mekong River Basin* (Brill Nijhoff, 2014) 37.

421 *Id.* at 38.

Vietnam enacted the Law on Water Resources in 1998, which, according to Hirsch and Jensen, provides the "framework for a permit system, ... [but] little guidance on sustainable water management."[422] Paragraph 1 of Article 22 of the Law on Water Resources states that:

> [Organizations and individuals] have the right to exploit and use water resource for purpose of living, agricultural, forestry and industrial production, mining, electricity generating, water transport, aquaculture, sea fishery, salt making, sport, recreation, tourism, medicine, health rehabilitation, scientific research and other purposes as stipulated by this Law and other provisions of law.

Paragraph 1 of Article 24 provides that "[o]rganizations and individuals that exploit and use water resources must get permission from the competent State agencies except [for] the cases stipulated in Clause 2 of this Article." Like Thailand, Vietnam suffers from problems with agency coordination and cooperation. Basically, two ministries, Natural Resources and the Environment (water quality) and Agriculture and Rural Development (water quantity), have "overlapping responsibilities ... [and] compete for jurisdiction over water."[423] There are insufficient policies in place to ensure water quality, water quantity, public health, and ecosystem health are considered thoroughly in the decision-making process of these ministries or that they work together to ensure comprehensive holistic water resources management. No provision in the Law on Water Resources makes a single reference to the MRC. The Vietnam Mekong Committee is established under the auspices of the Ministry of Agriculture and Rural Development.[424]

Hydropower development on Mekong tributaries is a key priority for Vietnam, as evidenced by such projects as the "Sesan cascade" (Yali Falls, Dak Bla, Plei Krong, Upper Kon Tum, Sesan 3, Sesan 3A, Sesan 4 and Sesan 4A hydroprojects).[425] Vietnam's increased demand for energy from a burgeoning population and growing economy has spurred a hydropower sector in that country, seemingly in competition with the Lao PDR, which has its own

---

422 Hirsch and Jensen, *supra* note 180, at 36.

423 Alistair Rieu-Clarke and Geoffrey Gooch, "Governing the Tributaries of the Mekong–The Contribution of International Law and Institutions to Enhancing Equitable Cooperation Over the Sesan" (2010) 22(2) *Pac. McGeorge Global Bus. and Dev. L.J.* 193, 216.

424 *See* Government Decree on Functions, tasks, authorities and organizational structure of Ministry of Agriculture and Rural Development, Vietnam (2003).

425 Rieu-Clarke and Gooch, *supra* note 423, at 201.

aspirations to become the battery of Southeast Asia. According to Rieu-Clarke and Gooch, growth in demand for power between 2001 and 2020 is forecast to be ten to eleven percent annually fueled by an economy with growth of eight percent until 2010.[426] But perhaps the most pressing national interest is the need to protect agriculture in the state's "rice bowl," the expansive Mekong delta, and to maintain dry season flows to counter saltwater intrusion into the delta. It would seem that the MRC could play a role in the emerging conflict challenges between the hydropower and irrigation interests by promoting basin-wide, integrated watershed management; yet so far this has not been the case.

In 1996, the Lao PDR enacted the Water Resources Law, but nevertheless water resources management remains sectoral and in the hands of numerous agencies. The Water Resources Law does not mention the current Mekong legal regime. A separate policy framework, the so-called Policy on Water and Water Resources, provides guidance on policy development but "is not legally binding."[427] The Water Resources Coordination Committee is noteworthy because it includes the Lao NMC, established by decree in 1999.[428] The domestic legal regime of water in the Lao PDR basically classifies water uses as small-, middle- and large-scale, with responsibilities for granting permission shared between national and local governments. It comes as no surprise that the majority of water licenses issued to date have been to the hydropower sector which provides a major source of revenue through sales of power to Thailand.[429] Indeed, hydropower is the centerpiece of the Lao PDR's water management program given that the country has roughly half of the hydropower potential in the LMRB.[430] Megaprojects such as the Nam Theun, Nam Theun 2, and Nam Leuk dams have all been developed on Mekong tributaries in the Lao PDR.[431]

According to the Water Resources Law, hydroelectric projects should be planned "with due consideration for environmental protection, flood prevention, water supply, irrigation, navigation, [and] fisheries";[432] yet the law does not establish sufficient institutional processes to actually consider many

426 *Id.* at 200 (internal citations omitted).

427 Hirsch and Jensen, *supra* note 180, at 40.

428 *Id.*

429 McCaffrey, *supra* note 9, at 287; Dinar et al., *supra* note 26, at 234.

430 Dinar et al., *supra* note 26, at 234.

431 McCaffrey, *supra* note 9, at 287.

432 *See Law on Water and Water Resources* (Lao People's Democratic Republic) National Assembly, 11 October 1996, art. 25.

of these needs during the planning phase.[433] The Lao PDR has developed a National Water Resources Sector Strategy and Action Plan, but Gao notes that this does little to establish impactful consideration of the aforementioned issues when designing and locating potential hydropower projects,[434] let alone the consideration of the needs of other stakeholders in the Mekong region.

Cambodia remains one of the poorest countries in the world and is in a state of "nation building," factors which contribute to sluggish development and poor coordination of water law and policy.[435] Water law and policy are being developed by the Ministry of Water Resources and Meteorology, the lead water agency in Cambodia. Cambodia passed the landmark Law on Water Resources Management in 2007, yet the implementation of its IWRM-based provisions remains to be seen.[436] Specifically, the law purports to employ IWRM by establishing the Ministry of Water Resources and Meteorology as the lead in water resources planning; requiring collaboration and data sharing between the Ministry, other agencies, and local stakeholders; transparency and disclosure of decision-making activities; and a provision prioritizing sustainable development of Cambodia's water resources.[437] However, according to Gao, "Cambodia represents a classic illustration of the difficulties in applying the ideologies of IWRM in the context of an entrenched water regime," citing the nation's relatively recent political turbulence, and the unfamiliarity of its citizens and officials with access to information and civic engagement.[438] Strengthening agency policy through rulemaking processes could help Cambodian agencies become more effective and provide for comprehensive and holistic water resources management. A water law and policy project similar to the American Bar Association's Cambodia Law and Democracy Project[439] could benefit this impoverished state.

Cambodia's national interests focus on the fisheries of the Tonle Sap system and irrigation. Under Article 5(A) of the 1995 Mekong Agreement., the Tonle

433 Gao, *supra* note 420, at 39.

434 *Id.*

435 *See generally,* Gottesman, *supra* note 35, describing the current state of nation building in Cambodia from his vantage point as resident liaison and deputy director of the American Bar Association Cambodia Law and Democracy Project.

436 Gao, *supra* note 420, at 40.

437 *Law on Water Resources Management* (Kingdom of Cambodia), National Assembly, 29 June 2007, arts. 1, 4, 7, 8, 19.

438 Gao, *supra* note 420, at 41.

439 See American Bar Association, Cambodian Law and Democracy Project, "Needs Assessment Report," Phnom Penh Office (June 1995) available at <http://pdf.usaid.gov/pdf_docs/PNABX357.pdf> (accessed 14 June 2017).

Sap is considered a tributary, and accordingly, Cambodia maintains priority water use rights and control over it. This designation befits the hydrology of the Mekong River system, although the Tonle Sap receives most of its water from mainstream wet season flows.

Simply stated, water resources management in the LMRB states is sectoral and often disjointed. Two of the principal reasons that this is so is the disparate treatment of tributaries in the current Mekong legal regime and the emphasis on national sovereignty. Unless tributaries are demarginalized in the current legal framework, then the indictment of posterity will rightly point its finger at the LMRB states who have not been released from their obligations, but given an opportunity to reform their policies, domestic laws, and the 1995 Mekong Agreement to comport with customary international law.

2.2.3 *The Role of International Water Law*

At this point in the analysis of the Mekong legal regime, a review of the fundamental principles of international water law will benefit this discourse. Customary international water law played an important role in framing the provisions of the 1995 Mekong Agreement.[440] According to McCaffrey, "[i]nternational water law seeks to reconcile the sovereign rights of states sharing freshwater resources so as to optimize benefits for all, while ensuring adequate protection of the watercourse."[441] There are two basic forms of international water law: (1) treaty law and (2) customary international law.[442] Where no treaty exists concerning shared waters, customary international law governs states' rights and obligations.[443]

The three main principles of customary international law on watercourses are equitable and reasonable utilization, prevention of significant harm, and prior notification—rules which "greatly restrict the scope for purely unilateral action."[444] Noteworthy is that "[s]tates planning projects on international watercourses, *including tributary streams* ..., must account for" these rules.[445] McCaffrey sets forth these three rules as follows:

---

440 Radosevich and Olson, *supra* note 316, at 1.

441 Stephen McCaffrey, "Establishing Rights—The Law of International Waters" in Aaron T. Wolf (ed.), *Sharing Water, Sharing Benefits: Working Towards Effective Transboundary Water Resources Management* (A Graduate/Professional Skills-Building Workbook, adapted from the World Bank International Waters Course, 2010) 192–193.

442 *Id.*

443 *Id.*

444 *Id.*

445 *Id.* (emphasis added).

1. *Equitable and reasonable utilization*: Shared water must be used in a manner that is equitable and reasonable vis-à-vis co-riparian states. What is "equitable and reasonable utilization" is determined case-by-case, taking into consideration all relevant factors based upon both natural and human-related phenomena.
2. *Prevention of significant harm*: Countries must do their best to prevent uses within their territories from causing significant harm to other states. Probably most controversial issue in international water law-relationship between equitable utilization and prevention of significant harm. The [1997] UN Convention seems to suggest that one state's use can cause some harm to another state and still be justified as equitable.
3. *Prior notification*: [A] state must notify other states of planned activities that may adversely affect those other states. Potentially affected states must be permitted to comment on and consult with the notifying state concerning the plans.[446]

Other important concepts, principles and customary laws "that were fundamental to discussion on ... [the 1995] Mekong Agreement" include the sovereignty equality of states,[447] the principle of international waters (watercourse concept),[448] and the duty to cooperate.[449]

State sovereignty implies powerful concepts of jurisdiction, territory, control, and independence. It has two meanings in connection with international watercourses: sovereignty over a watercourse and the sovereign equality to share in the benefits of a watercourse.[450] McCaffrey explains how these two meanings of sovereignty fit within international watercourse law's centerpiece-equitable and reasonable utilization:

> It is clear ... that there is overwhelming support for the doctrine of equitable utilization as a general, guiding principle of law for the determination of the rights of States in respect of the non-navigational uses of international watercourses.[451]

446 *Id.* (emphasis in original).

447 Radosevich and Olson, *supra* note 316, at 1.

448 *Id.* at 2.

449 *Id.* at 3.

450 Margaret J. Vick, *Integrated Water Law: Local to International* (unpublished JSD dissertation, McGeorge School of Law, University of the Pacific, September 2009) 109 (on file with author).

451 Stephen C. McCaffrey, (Special Rapporteur), *Second Report on the Law of the Non-Navigational Uses of International Watercourses*, U.N. Doc. A/CN.4/399 (1986) Par. 170 [hereinafter McCaffrey, *Second Report*].

> The bedrock upon which the doctrine of equitable utilization is founded is the fundamental principle represented by the maxim *sic utere tuo ut alienum non laedas.* As seen above, this maxim is a generally accepted principle of law governing the relations between States. In the context of the use of a watercourse which separates or traverses two or more States, this means that one of those States may not use or permit the use of the watercourse in such a way as to cause injury to the other(s). Thus the States are referred to as having "equal" or, perhaps more accurately, "correlative" rights in respect of use of the watercourse, a concept which finds expression in the doctrine of limited territorial sovereignty: *a State has the sovereign right to make whatever use it wishes of waters within its territory, but that right is limited by the duty not to cause injury to other States.*[452]

The *sic utere* maxim in the context of equitable and reasonable utilization means denial of another state's equitable share, not simply factual harm. Sovereign equality to share in the benefits of a watercourse is, then, a fundamental precept of equitable and reasonable utilization. However, no state is entitled to share in the benefits of a watercourse to the extent that it interferes with another state's equitable and reasonable share.[453]

Sovereignty over a watercourse naturally evokes the concept of a state's right to develop a watercourse in terms of a national agenda. There is an inherent tension between state territorial sovereignty and development of transboundary tributary water resources. "States have, in accordance with the Charter of the United Nations and the principles of international law, the sovereign right to exploit their own resources pursuant to their own environmental policies."[454]

This concept was reinforced and expanded in Principle 2 of the Rio Declaration which states:

> States have, in accordance with the Charter of the United Nations and the principles of international law, the sovereign right to exploit their own resources pursuant to their own environmental and developmental policies, and the responsibility to ensure that activities within their

452 *Id.* at Par. 171 (citations omitted) (emphasis added).

453 *Id.* at Par. 75.

454 *Declaration of the United Nations Conference on the Human Environment* (Stockholm Declaration), Principle 21, 11 ILM 1416 (16 June 1972).

> jurisdiction or control do not cause damage to the environment of other States or of areas beyond the limits of national jurisdiction.[455]

The right to develop includes a state's right to "the exercise of [the] … inalienable right to full sovereignty over all … [its] natural wealth and resources."[456] The oft cited General Assembly (GA) Resolution 1803(XVII) provides that "[t]he right of peoples and nations to permanent sovereignty over their natural wealth and resources must be exercised in the interest of their national development and of the well-being of the people of the State concerned."[457] Sovereignty over natural resources is a principle of customary international law:

> The principle of "permanent sovereignty over natural resources" as the right of peoples and nations to use and dispose of the natural resources in their territories in the interest of their national development and well-being was established by the General Assembly in its resolution 1803 (XVII) of December 1962 …. While the legal nature of the core principle of "permanent sovereignty over natural resources", as a corollary to the principle of territorial sovereignty or the right of self-determination, is indisputably part of customary international law, its exact legal scope and implications are still debatable.[458]

Vick notes that "the question remains as to … [permanent sovereignty's] applicability to an international watercourse."[459] "The legal theories related to transboundary watercourses of absolute territorial sovereignty and absolute territorial integrity are comparable to the policy of permanent sovereignty over natural resources."[460] Accordingly, the four theoretical bases of international watercourse law should also be mentioned in any discourse on transboundary tributary watercourse law: (1) absolute territorial sovereignty or "the Harmon

455 Rio Declaration, *supra* note 220, A/Conf.151/26, 31 ILM 874 (1992).

456 Declaration on the Right to Development A/RES/41/128 adopted by the General Assembly 4 December 1986.

457 UN General Assembly resolution 1803 (XVII) *Permanent Sovereignty over Natural Resources*, 14 December 1962.

458 Letter dated 29 January 2002 from the Under-Secretary-General for Legal Affairs, the Legal Counsel, addressed to the President of the Security Council, 12 February 2002, S/2002/161 (regarding contracts signed by Moroccan authorities with foreign companies for the exploration of mineral resources in Western Sahara).

459 Vick, *supra* note 450, at 110.

460 *Id.* at 110–111.

Doctrine"; (2) absolute territorial integrity; (3) limited territorial sovereignty; and (4) community of interest.[461]

The first theory, absolute territorial sovereignty, is an extreme concept that infers that a state may do as it pleases with its watercourses without any consideration of the adverse effects its actions may have on other states.[462] It is based on the theory that a state has an absolute right to use the water that flows through its territory.[463] The foundation of this theory relates back to the 1895 dispute over the Rio Grande, when US Attorney General Judson Harmon denied any responsibility for the adverse effects on Mexico of US upstream water diversions on the Rio Grande.[464] Harmon stated: "the rules, principles and precedents of international law impose no liability or obligation upon the United States."[465] He concluded that claims by Mexico were "entirely inconsistent with the sovereignty of the United States over its national domain."[466] This theory favors upstream riparians since it absolves them from any responsibility for harm to downstream riparians for any adverse consequences of the use of a watercourse by the former. The theory is generally rejected as inconsistent with the interdependence that characterizes modern state practice and did not figure into the ultimate resolution of the dispute that gave rise to it.

The antithesis of absolute territorial sovereignty is the theory of absolute territorial integrity, which provides that "the upstream state may do nothing that might affect the natural flow of the water into the downstream state."[467] This theory has the potential to severely impact upstream states that have been slower in developing their water resources than their downstream neighbors.[468] Its operative effect would clearly prohibit the alteration of flow regimes and flood pulses or diminution in water quantity or quality by upstream riparians at the expense of downstream ones. The absolute territorial integrity doctrine has been cited by downstream states such as Egypt in the Nile dispute and Pakistan in the Indus conflict.[469] Like the Harmon Doctrine,

---

461 *See generally,* McCaffrey, *supra* note 9, at 111–133 (describing the theoretical basis of international watercourse law and an analysis of the four principal theories).

462 Stephen C. McCaffrey, "The Harmon Doctrine One Hundred Years Later: Buried, Not Praised" (1996) 36 *Nat. Resources J.* 549, 551 [hereinafter McCaffrey (1996)].

463 *Id.*

464 *Id.*

465 Patricia Wouters, (ed.), *International Water Law: Selected Writings of Professor Charles B. Bourne* (Kluwer Law International, 1997) 38.

466 *Id.*

467 McCaffrey, *supra* note 9, at 126.

468 *Id.*

469 *Id.* at 128–129.

absolute territorial integrity is much too "factually myopic and 'legally anarchic'" for use in transnational practice.[470]

According to McCaffrey, "[t]he doctrine of limited territorial sovereignty is probably the prevailing theory of international watercourse law and obligations today.[471] According to this doctrine, "[t]he sovereignty of a state over its territory is said to be 'limited' by the obligation not to use that territory in such a way as to cause significant harm to other states."[472] This theory is a balance between the two aforementioned extreme concepts. The weight of state practice, along with modern legal and hydropolicy literature and decisions, points toward "the proposition that states sharing an international watercourse have rights to the use of its waters, that those rights are, in principle, equal, and that accordingly, each state must respect the rights of the other."[473] This doctrine of limited territorial sovereignty was strongly supported by the International Court of Justice (ICJ) in the *Case concerning the Gabcíkovo-Nagymaros Project* (Hungary/Slovakia) (*Gabcíkovo-Nagymaros* case)[474] in which the principle of equitable utilization was dispositive of water use issues on the Danube. Thus, an "equality of right" exists in an international watercourse which endorses that a river and its tributaries are a common resource to be shared between the riparian states. Although the right to share is equal, this does not infer a right to an equal amount of water for each riparian. Equitable use is, however, based on relevant factors and circumstances.[475]

Equitable utilization is derivative of the concept of equitable apportionment fashioned by the U.S. Supreme Court in *Kansas v. Colorado*,[476] a case

---

470 *Id.* at 133 (citation omitted).

471 *Id.* at 135.

472 *Id.*

473 *Id.* at 135–136 (citations omitted).

474 Judgment of 25 September 1997, 1997 ICJ 7; *reprinted in* 37 ILM 162 (1998), also available on the ICJ web site, <www.icj-cij.org/docket/files/92/7375.pdf> (accessed 14 June 2017) [hereinafter *Gabcikovo-Nagymaros* case].

475 *See* UN Watercourses Convention, *supra* note 18, arts. 5–6.

476 206 U.S. 46 (1907). This case established the doctrine of equitable apportionment of interstate water resources. Relying on riparian rights and the natural flow doctrine, Kansas attempted to restrict the use of the Arkansas River upstream in Colorado, a prior appropriation or "first in time, first in right" state. The US Supreme Court acknowledged that economic interests in Kansas might be harmed by Colorado's use of the Arkansas River, but nevertheless ruled that Colorado was entitled to a share of those waters. The Court considered Kansas and Colorado water laws (riparian rights versus prior appropriation) and concluded that when state water laws and policies were in conflict, that equity would control and the interests of the competing states would be balanced. Colorado prevailed because Kansas was unable to prove to the Court's satisfaction that the latter state's interest would be harmed by water diversions in the former state.

concerning mutual assertions of state sovereignty that replicates the legal dynamics played out in international watercourse conflicts. In this seminal case, a battle over the Arkansas River ensued when Kansas, the downstream riparian rights state that developed water resources first for agricultural development, power production, manufacturing, and milling, challenged the upstream prior appropriation state, Colorado, over new water uses in the latter state for irrigation. Colorado asserted sovereignty "to appropriate all the waters of this stream for the purposes of irrigating its soil and making more valuable its territory."[477] Kansas asserted "that the flowing water in the Arkansas must, in accordance with the extreme doctrine of the common law of England, be left to flow as it was wont to flow, no portion of it being appropriated in Colorado for the purposes of irrigation."[478] The Court rejected both contentions as "extreme,"[479] stating that "[o]ne cardinal rule, underlying all the relations of the states to each, is that of equality of right."[480] The Court balanced the benefits to Colorado versus the harm to Kansas[481] and decided that neither state had exclusive control over waters of the Arkansas River, but rather both states were entitled to an equitable apportionment and accordingly, the lawsuit was dismissed.[482]

Article 4 of the 1995 Mekong Agreement explicitly declares that the LMRB states are "[t]o cooperate on the basis of sovereign equality and territorial integrity in the utilization and protection of the water resources of the Mekong River Basin." The approach in the Mekong legal regime, as far back as the inception of the Water Charter in 1971, has been that tributaries remained under a national sovereignty agenda. The LMRB states basically attempted to vest themselves with full sovereign jurisdiction over planned measures and activities on tributaries within their territories, seemingly ignoring that customary international law takes over where any treaty leaves off. In contrast with the mainstream, the legal regime of tributaries is based on hydrosovereignty, irrespective of the transboundary nature of some of these important arteries. Such international tributaries should normally follow the same regime applicable to the mainstream. Noteworthy is that "[t]he 1997 UN [Watercourses] Convention does not contain a corresponding provision on sovereignty."[483]

477 *Id.* at 98.

478 *Id.*

479 *Id.*

480 *Id.* at 97.

481 *Id.* at 100.

482 *Id.* at 117–118.

483 Margaret J. Vick, "International Water Law and Sovereignty: A Discussion of the ILC Draft Articles on the Law of Transboundary Aquifers" (2008) 21 *Pac. McGeorge Global Bus. and Dev. L.J.* 191, 208.

Indeed, this comprehensive framework instrument and its antecedents downplay the concept.[484]

While the duty to cooperate may seemingly infer a need for a forum and an organization, the UN Watercourses Convention does not require watercourse states to establish a joint basin institution, although it does encourage one when appropriate.[485] However, the UN Watercourses Convention does require the exchange, on a regular basis, of "available data and information on the condition of the watercourse, in particular that of a hydrological, meteorological, hydrogeological, and ecological nature and related to the water quality as well as related forecasts."[486] Data exchange is anticipated to foster a continuing spirit of cooperation and allows watercourse states to practice due diligence in their planned measures.

One of the important currents in international watercourse law is the theory of a "community of interest," which has emerged since about 1929 but is couched in antiquity.[487] In the 1929 case concerning the *Territorial Jurisdiction of the International Commission of the River Oder* (*River Oder* case),[488] the Permanent Court of International Justice (PCIJ) was asked: (1) whether the jurisdiction of the International Commission of the River Oder and navigation rights of states other than Poland extended under the provisions of the Treaty of Versailles to the sections of tributaries of the Oder, the Warthe (Warta), or the Netze (Notec)[489] flowing through Poland, and (2) what principle was appropriate for determining the upstream limits of the Commission's jurisdiction. Based on an interpretation of Article 331 derived from the "principles governing international fluvial law in general,"[490] the PCIJ stated that there

484 *See* Stephen M. Schwebel, (Special Rapporteur), *Third Report on the Law of the Non-Navigational Uses of International Watercourses,* A/CN.4/348 (1 December 1981) para. 45; McCaffrey, *Second Report, supra* note 451, para. 165.

485 UN Watercourses Convention, *supra* note 18, art. 8.

486 *Id.*, art. 9(1).

487 McCaffrey, *supra* note 9, at 147 and 150. According to McCaffrey, the theory of a community of interest "has antecedents in Roman [water] law" wherein perennial streams are considered to be "common or public, *rei publicae jure gentium.*" *See* Institutes of Justinian, lib II, titl., secs. 2 and 5.

488 Judgment No. 16, 10 September 1929, PCIJ Ser. A No. 23, pp. 5–46. Reproduced in Manley O. Hudson ed., WORLD COURT REPORTS, vol. 2, p. 609 (1969) [hereinafter the *River Oder* case].

489 The Netze (Notec) and Warthe (Warta) rise in Poland, flow a long way through that country, from the German-Polish frontier and ultimately flow into Germany where the Netze (Notec) flows into the Warthe (Warta) before the latter joins the Oder.

490 Article 331 of the Treaty of Versailles provides: "The following rivers are declared international: ... the Oder (*Odra*) from its confluence with the Oppa; ... and all navigable parts of these river systems which naturally provide more than one State with access to the sea

was a "community of interest" in transboundary watercourses which "becomes the basis of a common legal right, the essential features of which are the perfect equality of all riparian States in the use of the whole course of the river and the exclusion of any preferential privilege of any one riparian State in relation to the others."[491] McCaffrey notes that "a community of interest in the water is created by the natural, physical unity of a watercourse."[492] Holding that the Commission's jurisdiction extended to both tributaries, the Court handed down the community of interest concept that today has evolved into a theoretical basis for the law of international watercourses. This case is reviewed in Section 3.2.3 of this book.

"[S]everal recent agreements and draft treaties reflect what may be regarded as a community-of-interest approach."[493] Noteworthy in the context of the

…." The "regime of internationalization" of the Oder arose from Articles 332 to 337 of the Treaty of Versailles. PCIJ Ser. A No. 23, at 23 and 26.

491 PCIJ Ser. A, No. 23, *supra* note 488, at 27–28.

492 McCaffrey, *supra* note 9, at 148 (citing Juraj Andrassy, "Les Relations Internationales de Voisinage", (1951-II) 79 *Recueil Des Cours* 77–181, at 104: "C'est donc l'unite physique ou naturale qui cree la communauté d' interest.").

493 McCaffrey, *supra* note 9, at 152–154. See the Protocol on Shared Watercourse Systems in the Southern African Development Community Region, 28 August 1995, entered into force 29 December 1998, art. 1(1) [hereinafter SADC Protocol] stands out as exemplary in this context *in* Treaties Concerning the Non-Navigational Uses of International Watercourses, Africa, FAO Legislative Study 61, p. 146, at pp. 147–148 (1997). The agreement was prepared by members of SADC, including Angola, Botswana, Lesotho, Malawi, Mozambique, Namibia, Swaziland, Tanzania, Zambia and Zimbabwe, and was also signed by South Africa. The SADC Protocol includes the Zambezi, Okavango, Orange, Pungue, Cunene and Save rivers. The SADC Protocol was revised on August 7, 2000 as the Revised Protocol Shared Watercourses in the Southern African Development Community [hereinafter Revised SADC Protocol]. The Revised SADC Protocol takes into account the principles in the 1997 UN Convention and uses the term "shared watercourse" which is defined in art. 1.1 as a "watercourse passing through or forming the border between two or more Watercourse States." It asserts a determination to "foster closer cooperation for judicious, sustainable and coordinated management, protection and utilization of shared watercourses and advance the SADC agenda of regional integration and poverty alleviation; A similar approach is found in the Agreement of September 14, 1992, between Namibia and South Africa on the Establishment of a Permanent Water Commission, 32 ILM 1147 (1993), art. 1(2). The term "common" is included in the title of the Agreement of July 18, 1990, between Nigeria and Niger concerning the Equitable Sharing in the Development, Conservation and Use of Their Common Water Resources. Copy on file with the Food and Agriculture Organization of the United Nations (FAO) in Rome. This treaty uses the term "shared river basins." *See, e.g.*, arts. 1 and 2. ; Finally, a community of interest approach is reflected in the Agreement between Bolivia and Peru concerning a Preliminary Economic Study of

community of interest theory is the use of the phrase "community of Mekong nations" in the Mekong legal regime.[494] Paragraph 6 of the Preamble of the 1995 Mekong Agreement affirms a spirit of cooperation:

> [T]o promote or assist in the promotion of interdependent sub-regional growth and cooperation among *the community of Mekong nations*, taking into account the regional benefits that could be derived and/or detriments that could be avoided or mitigated from activities within the Mekong River Basin undertaken by this framework of cooperation. (emphasis added).

This paragraph, interpreted broadly, has far-reaching implications for tributaries in the Mekong legal regime given that these indicia of some type of community of interest or interdependency approach to growth and cooperation mentioned in the Preamble are declared part of the 1995 Mekong Agreement by Article 38.[495] This begs the question of whether a community of interest approach is a corollary of the "community of Mekong nations."

---

the Joint Utilization of Lake Titicaca of 19, 1957, LEGISLATIVE TEXTS, Treaty No. 45, p. 168, ratified by Peru in 1957 and by Bolivia in 1987. un, *International Rivers and Lakes Newsletter*, No. 23, at 3 (June 1995). McCaffrey notes that art. 1 of that agreement refers expressly to "the fact that the two countries have joint, indivisible and exclusive ownership over the waters of Lake Titicaca." In exchanges of notes of 12 December 1992 and 18 May 1993, the parties agreed to establish a Binational Authority for the implementation of the Binational Master Plan of the Titicaca-Desaguadero-Poopo-Salar de Copaisa System. *See* un, *International Rivers and Lakes Newsletter*, No. 23, at 4 (June 1995). McCaffrey explains that "[j]oint planning and implementation are hallmarks of a community-of-interest approach to the shared use and development of international watercourses." Also noteworthy is that the foregoing agreements involve developing countries or have a developing country as one party.

494 1995 Mekong Agreement, *supra* note 1, para. 6.

495 1995 Mekong Agreement, *supra* note 1, art. 38. Article 38 of the 1995 Mekong Agreement, "Scope of Agreement," provides: "This Agreement shall consist of *the Preamble* and all provisions thereafter and amendments thereto, the Annexes, and all other agreements entered into by the Parties under this Agreement. Parties may enter into bi- or multilateral special agreements or arrangements for implementation and management of any programs and projects to be undertaken within the framework of this Agreement, which agreements shall not be in conflict with this Agreement and shall not confer any rights or obligations upon the parties not signatories thereto, except as otherwise conferred under this Agreement." (emphasis added).

2.2.3(a) Beginnings: The 1966 Helsinki Rules

The Helsinki Rules on the Uses of the Waters of International Rivers (1966 Helsinki Rules),[496] adopted by the International Law Association (ILA) in 1966, opened new vistas as the first attempt at codification of international water law covering both "navigational and nonnavigational activities."[497] Drafted by the ILA's Committee on the Uses of the Waters of International Rivers, the Rules were prepared "to clarify and restate existing international law as it applies to the rights of states to utilize the waters of an international drainage basin."[498] McCaffrey calls them a "monumental work … [and] important precursors of the ILC's [International Law Commission] draft articles and thus of the [UN Watercourses] Convention."[499] Indeed the Rules championed the concept of an international drainage basin,[500] which insightfully included groundwater, and the principles of equitable and reasonable sharing in beneficial water uses, commonly referred to as equitable utilization.[501]

---

496 The 1966 Helsinki Rules, *supra* note 260, were revised and updated in 2004 with the adoption by the ILA of the "Berlin Rules."

497 Lucius Caflisch, "Regulation of the Uses of International Waterways: The Contribution of the United Nations" in Martin Ira Glassner (ed.), *The United Nations at Work* (Greenwood, 1998) 3–35, 4.

498 1966 Helsinki Rules, *supra* note 260, at 1.

499 McCaffrey *supra* note 9, at 381.

500 1966 Helsinki Rules, *supra* note 260, art. II. Article II of the 1966 Helsinki Rules provides:"An international drainage basin is a geographical area extending over two or more States determined by the watershed limits of the system of waters, including surface and underground waters, flowing into a common terminus."

501 1966 Helsinki Rules, *supra* note 260, art. V. Article V of the 1966 Helsinki Rules provides: "I. What is a reasonable and equitable share within the meaning of article IV to be determined in the light of all the relevant factors in each particular case.

II. Relevant factors which are to be considered include, but are not limited to:

1. The geography of the basin, including in particular the extent of the drainage area in the territory of each basin State;
2. The hydrology of the basin, including in particular the contribution of water by each basin State;
3. The climate affecting the basin;
4. The past utilization of the waters of the basin, including in particular existing utilization;
5. The economic and social needs of each basin State;
6. The population dependent on the waters of the basin in each basin State;
7. The comparative costs of alternative means of satisfying the economic and social needs of each basin State;
8. The availability of other resources;
9. The avoidance of unnecessary waste in the utilization of waters of the basin;

The Rules also contain provisions on pollution control and prevention,[502] navigation,[503] and dispute settlement.[504] The 1966 Helsinki Rules constitute an important source of soft law, contribute significantly to international water resources law, especially the drainage basin concept, and "reflect many principles and trends that later found expression in the [more comprehensive] UN [Watercourses] Convention."[505]

The Rules' significance to tributary watercourse law in the LMRB is that "they take a holistic, basin approach."[506] According to the ILA's commentary to Article II an international drainage basin is an indivisible hydrologic unit.[507] McCaffrey explains the importance of this commentary to tributary watercourse law:

> This commentary leaves no doubt that the rules of international law stated by the Helsinki Rules apply not only to the main stem of a river, or to portions of a stream forming a boundary, but to *tributaries* of an international watercourse as well.
>
> As observed in a later report of the ILA's Committee on International Water Resources Law, the adoption of the drainage basin approach to the codification of the law of international watercourses represented a significant step forward. For the first time it confirmed in terms of juridical

---

10. The practicability of compensation to one or more of the co-basin States as a means of adjusting conflicts among uses; and

11. The degree to which the needs of a basin State may be satisfied, without causing substantial injury to a co-basin State.

III. The weight to be given to each factor is to be determined by its importance in comparison with that of other relevant factors. In determining what is reasonable and equitable share, all relevant factors are to be considered together and a conclusion reached on the basis of the whole."

502 1966 Helsinki Rules, *supra* note 260, Chapter 3.

503 1966 Helsinki Rules, *supra* note 260, Chapter 4.

504 1966 Helsinki Rules, *supra* note 260, Chapter 6.

505 McCaffrey *supra* note 9, at 381.

506 *Id.*

507 1966 Helsinki Rules, *supra* note 260, at 8. The comments state:"With the relatively recent multi-use development of international rivers, the concern is no longer limited to the navigable portion of the international river, but rather encompasses all waters included in the entire system comprising the international drainage basin. [A] State, although not riparian to the principal stream of the basin, may nevertheless supply substantial quantities of water to that stream; such a State thus is in a position to interfere with the supply of water through action with respect to the water flowing within its own territory.'

> area what naturalists, engineers and economists had previously accepted: that the basin, as a physical whole and not only the river or the waters, must be the object of legal regulation.[508]

Indeed, over 50 years ago in his influential work on river basins and law, Teclaff recognized that "the waters which the river basin receives ... form an interconnected system which is capable of transmitting within itself any disturbance caused by changes affecting water in any part of the basin."[509] McCaffrey aptly notes that "[t]he fact that the waters within a drainage basin are interconnected *makes the basin the ideal unit to which to apply a legal regime*."[510]

In the Preamble and throughout the 1995 Mekong Agreement, the treaty refers to the Mekong River "Basin," a term undefined in the four corners of the document. The Agreement, then, purports to adopt a basin-wide approach and does not deal solely with just the Mekong River. Standard and geological dictionaries define a basin as consisting of the entire area drained by a "river and *its tributaries*."[511] Article 31(1) of the 1969 Vienna Convention on the Law of Treaties requires that a treaty be interpreted "in good faith in accordance with the ordinary meaning to be given to the terms of the treaty in their context and in the light of its object and purpose."[512] Consequently, the use of the term "basin" throughout the 1995 Mekong Agreement raises the implication of an indivisible hydrologic unit. Accordingly, it seems reasonable under the principles of legal interpretation and customary usage to interpret the word "basin" in the 1995 Mekong Agreement to include all tributaries. Several commentators have reached similar conclusions.[513] Wohlwend found that:

---

508 Stephen C. McCaffrey, "International Organizations and the Holistic Approach to Water Problems" (1991) 31 *Nat. Resources J.* 139, 143 (internal citations omitted) (emphasis added) [hereinafter McCaffrey (1991)].

509 Ludwik A. Teclaff, *The River Basin in History and Law* (The Hague/Martinus Nijhoff, 1967) 14.

510 McCaffrey (1991), *supra* note 508, at 144. (emphasis added).

511 *See Merriam-Webster Dictionary* <http://www.merriam-webster.com/dictionary/basin> (accessed 14 June 2017); Bates and Jackson, *supra* note 22, at 541. (emphasis added).

512 23 May 1969, in force 27 January 1980, 8 ILM 679 (1969) [hereinafter 1969 Vienna Convention].

513 *See* Bernard J. Wohlwend, *The Emerging Principles and Rules of International Water Resources Law* 9 (March 2001), available at <http://www.bjwconsult.com/Bibliography.htm> (accessed 14 June 2017); *See also,* Sergei Vinogradov et al., *Transforming Potential Conflict into Cooperation Potential: The Role of International Water Law* (UNESCO/IHP, 2003) 48, available at <http://unesdoc.unesco.org/images/0013/001332/133258e.pdf> (accessed 14 June 2017).

> If the Title, Preamble and most Articles of the Agreement refer to the Mekong River Basin, and although no mention is made of groundwater resources, the definition of the term "Proposed use" in Chapter II refers to the "river system", as does Article 7 on the prevention and cessation of harmful effects which refers to "the aquatic (eco-system) conditions, and ecological balance of the river system", from which it may be inferred that the Mekong Basin consists essentially of the aggregate of the mainstream *and tributary watersheds* and that groundwaters are to be taken into account as elements of the ecosystem comprised of the water and related resources of the Basin. This point of view is further corroborated by the definition of the Environment at Chapter II which is said to encompass "the conditions of water and land resources, air, flora and fauna that exist in a particular region" or a particular ecosystem.
>
> In the absence of a specific definition of the Basin and of a reference to the Mekong Delta as its common terminus, it may nevertheless be asserted that, even if falling short of it, the object of this Agreement is in substance the Drainage Basin as defined in the 1966 Helsinki Rules.[514]

Similarly, Vinogradov, et al., reasoned that:

> The 1995 Agreement on the Cooperation for the Sustainable Development of the Mekong River Basin applies to the "water and related resources of the Mekong river basin." It also uses on one occasion the term "Mekong river system." However, neither of the terms is defined in the document.
>
> It follows from the text that the "system" comprises at least the mainstream of the Mekong river and its tributaries, including Tonle Sap.[515]

Rieu-Clarke and Gooch posit that the term "Mekong River Basin" is synonymous with that of "drainage basin" used in the 1966 Helsinki Rules.[516] They state that "[s]uch a definition, in its reference to a 'system of waters', would imply that the [1995 Mekong] Agreement *covers both the mainstream of the Mekong and its tributaries*."[517] The 1966 Helsinki Rules and the attendant commentary can thus be viewed as authoritative soft law in defining the term "basin" in the

514 Wohlwend, *supra* note 513, at 9 (emphasis added).

515 Vinogradov et al., *supra* note 513, at 48.

516 Rieu-Clarke and Gooch, *supra* note 423, at 208.

517 *Id.* (emphasis added).

1995 Mekong Agreement. Arguably, the 1966 Helsinki Rules provide a significant evidentiary foundation supporting inclusion of tributaries in the Mekong River Basin, and therefore, within its legal regime.

2.2.3(b) The Contribution of the UN Watercourses Convention

With the establishment of the United Nations in 1945, a new frontier for the development and codification of international law emerged. At its Plenary Meeting on 21 November 1959, the GA adopted Resolution 1401, Preliminary Studies on the Legal Problems Relating to the Utilization and Use of International Rivers, paving the way for development and codification of international watercourse law.[518] The Resolution pointedly highlighted the need for an evaluation of such an international legal regime. In 1970, the GA passed resolution 2669, Progressive Development and Codification of the Rules of International Law Relating to International Watercourses, and recommended that the ILC study the non-navigational uses of international watercourses with the aim of adopting a multilateral convention.[519] With the passage of these resolutions, the GA authorized the ILC to begin the deliberative work that ultimately led to the most significant contribution of that subsidiary organ to the law of international watercourses, the UN Watercourses Convention.

518 G.A. Res. 1401 (XIV), UN General Assembly Official Records (GAOR), 14th Sess., 842 Ple. Mtg., UN Doc. A/Res/1401 (XIV) (1959); See YBILC, vol. 2, Part 2, at 49 [1974]. G.A. Resolution 1401 provided: "That is desirable to initiate preliminary studies on the legal problems relating to the utilization and use of international rivers with a view to determining whether the subject is appropriate for codification.

The General Assembly then:

Requests the Secretary-General to prepare and circulate to Member States a report containing;

a) Information provided by Member States requesting their laws and legislation in force in the matter, and when necessary, a summary of such information;

b) A summary of existing bilateral and multilateral treaties;

c) A summary of decisions of international tribunals, including arbitration awards and;

d) A survey of studies made or being made by non-governmental organizations concerned with international law."

519 G.A. Res. 2669(XXV), UN GAOR, 25th Sess. UN Doc. Supp. No. 28, UN Doc. A/Res/2669 (XXV) (1970) *reprinted in* 9 ILM 1292 (1970). Resolution 2669 (XXV) was followed by Resolutions 2780 (XXVI), 2926 (XXVII), 3071 (XXVIII) and 3315 (XXIX). *See* YBILC, vol. 2, Part 1, at 150.

The UN Watercourses Convention entered into force on 17 August 2014 and codifies the essence of international watercourse law. It is the most comprehensive and important international agreement in the world's water resources spectrum and contains many principles which are broadly recognized to reflect customary international law.[520] The UN Watercourses Convention establishes a framework of guiding substantive and procedural principles, and its general approach has been noted with approval by the ICJ.[521]

In 1970, the ILC, as is customary, prepared a questionnaire concerning, *inter alia*, the drainage basin concept and submitted it to member states.[522] Results of the questionnaire revealed disparate views concerning the basis and approach for the work which "was mirrored in the ILC."[523] Not surprisingly, the division on this issue was largely along the upstream/downstream positions of the member states.[524] Those opposing the drainage basin approach were upstream states who preferred the expression "international river," a term, which McCaffrey explains, "[o]n its face, at least ... exclude[s] ... tributaries."[525]

In 1980, the ILC adopted six articles[526] employing the concept of the "international watercourse system."[527] The ILC's commentary to Article 1 explained that phrase:

---

520 *See* Stephen C. McCaffrey, "The Contribution of the UN Convention on the Law of the Non-Navigational Uses of International Watercourses" (2001) 1 *Int. J. Global Envtl. Issues* 250, 259 [hereinafter McCaffrey (2001)].

521 *See Gabcikovo-Nagymaros* case, *supra* note 474.

522 McCaffrey (1991), *supra* note 508, at 150–151. Three of the questions proffered to the member States are pertinent to tributary watercourse law and holistic water resources management. They are:

"(a) What would be the appropriate scope of the definition of an international watercourse, in a study of the legal aspects of fresh water uses, on the one hand and of fresh water pollution on the other hand?

(b) Is the geographical concept of an international drainage basin the appropriate basis for a study of the legal aspects of non-navigational uses of international watercourses?

(c) Is the geographical concept of an international drainage basin the appropriate basis for a study of the legal aspects of the pollution of international watercourses?" YBILC, vol. 2, Part 1, at 302, para. 17, UN Doc. A/CN.1/SER.A/1974.

523 McCaffrey (1991), *supra* note 508, at 151–153.

524 *Id.*, at 152–153.

525 *Id.*

526 *Id.*, at 153, footnote 84. *See* YBILC vol. 2, Part 2, at 110–136 (1980).

527 McCaffrey (1991), *supra* note 508, at 153–154.

> The Commission has selected this term because it gives the appropriate sense of dimension which characterizes an international watercourse. An international watercourse is not a pipe carrying water through the territory of two or more States. While its core is generally and rightly seen as the main stem of a river traversing or forming an international boundary, the international watercourse is something more, for it forms part of what may best be described as a "system"; it comprises components that embrace, or may embrace, not only rivers but other units *such as tributaries*, ... and groundwater, constituting by virtue of their physical relationship a unitary whole.[528]

McCaffrey concluded that "the concept ... comes very close to that of an 'international drainage basin' under the [1966] Helsinki Rules."[529] Also in 1980, the ILC used the term, "shared natural resource"[530] in Article 5, which points toward a holistic approach in terms of legal relativity.[531] In 1983–1984, concerns raised in the ILC about the use of both terms resulted in their elimination in the revised draft articles. The "shared natural resource" concept was replaced with a provision on equitable and reasonable use.[532] Later, the ILC decided to include the term "system" in 27 articles[533] it adopted in 1987, 1988, and 1990, but enclosed the term in brackets "indicating that no final decision [had] been reached on its use."[534] In a 1991 law journal article, Special Rapporteur

528 "Report of the Commission to the General Assembly on the work of its thirty-second session", *YBILC* (1980) vol. 2, Part 2, at 110–111 (emphasis added).

529 McCaffrey (1991), *supra* note 508, at 154.

530 See draft article 5, "Use of waters which constitute a shared natural resource," which provides:

"1. To the extent that the use of waters of an international watercourse system in the territory of one System. State affects the use of waters of that system in the territory of another system State, the waters are, for the purposes of the present articles, a shared natural resource.

2. Waters of an international watercourse system which constitute a shared natural resource shall be used by a system State in accordance with the present articles." YBILC, *supra* note 528, at 120 (1980).

531 YBILC, *supra* note 528, at 120 (1980); McCaffrey (1991), *supra* note 508, at 155.

532 See art. 6 *in* YBILC, vol. 2, Part 1, at 110 (1984).

533 McCaffrey (1991), *supra* note 508, at 157, footnote 106.

534 *Id.*, at 157–158. A footnote to art. 1, "[Use of terms]," reveals that the ILC "agreed to leave aside for the time being the question of article 1 (Use of terms) and that of the use of the

McCaffrey commented on the apparent departure of the ILC "from express recognition of the interdependence of the various components of hydrographic systems."[535] Noteworthy is the commentary to Article 2(b) of the 1994 ILC draft articles that states, "[t]he concept of a watercourse or river system is not a novel one. The expression has long been used in international agreements to refer to a river, *its tributaries* and related canals."[536] After more than two decades of work, in 1994 the ILC adopted this set of draft articles and attendant commentaries on the law of the non-navigational uses of international watercourses which formed the basis for the adoption of the UN Watercourses Convention.

The drafters of the UN Watercourses Convention did not adopt the drainage basin concept contained in the 1966 Helsinki Rules *per se* nor the "system" approach. Instead the term "watercourse" is used and "means a system of surface waters and groundwaters constituting by virtue of their physical relationship a unitary whole and normally flowing into a common terminus."[537] McCaffrey maintains that "a factually accurate definition of the term 'watercourse' must include not only the main surface water channel and the water contained therein, but also the other components of a watercourse system, in particular, *tributaries*, and groundwater."[538] According to McCaffrey, "the concept of a 'watercourse system' is conveyed appropriately by the expressions 'catchment', 'watershed' and '*drainage basin*', provided that these terms are understood to encompass subsurface water as well as surface water."[539]

Although watercourses transcend geopolitical boundaries, they are made "international" when superimposed with state boundaries. Thus, as defined in the UN Watercourses Convention, an "international watercourse system" is "a watercourse, parts of which are situated in different states."[540] The "parts" may consist of the mainstream, tributaries and interconnected groundwater.[541]

---

term "system" and to continue its work on the basis of the provisional working hypothesis accepted by the Commission ... in 1980." *See* YBILC, vol. 2, Part 2, at 25, n. 83 (1987).

535 McCaffrey (1991), *supra* note 508, at 158.

536 Draft articles on the *Law of the Non-Navigational Uses of International Watercourses* and commentaries, YBILC, vol. 2, Part 2, at 91, art. 2(b), commentary 8 (1994). (emphasis added).

537 UN Watercourses Convention, *supra* note 18, art. 2(a).

538 McCaffrey, *supra* note 9, at 34 (emphasis added).

539 *Id.* at 35. (emphasis added).

540 UN Watercourses Convention, *supra* note 18, art. 2(b).

541 McCaffrey, *supra* note 9, at 41.

The three main principles of customary international law on watercourses discussed above are codified in the UN Watercourses Convention: equitable and reasonable utilization,[542] prevention of significant harm,[543] and prior notification.[544] The centerpiece of the UN Watercourses Convention is Article 5 on equitable and reasonable utilization. In *The Donauversinkung Case*, the German Staatsgerichtshof noted that in interpreting this customary international norm:

> The application of this principle is governed by the circumstances of each particular case. The interests of the States in question must be weighed in an equitable manner against one another. One must consider not only the absolute injury caused to the neighboring State, but also the relation of the advantage gained by one injury cause to the other.[545]

---

542 UN Watercourses Convention, *supra* note 18, art. 5. Article 5, "Equitable and Reasonable Utilization and Participation," is the centerpiece of the Convention. It states:

"1. Watercourse States shall in their respective territories utilize an international watercourse in an equitable and reasonable manner. In particular, an international watercourse shall be used and developed by watercourse States with a view to attaining optimal and sustainable utilization thereof and benefits therefrom, taking into account the interests of the watercourse States concerned, consistent with adequate protection of the watercourse.

2. Watercourse States shall participate in the use, development and protection of an international watercourse in an equitable and reasonable manner. Such participation includes both the right to utilize the watercourse and the duty to cooperate in the protection and development thereof, as provided in the present Convention."

543 *Id.*, art. 7. Article. 7, "Obligation Not to Cause Significant Harm," provides:

"1. Watercourse States shall, in utilizing an international watercourse in their territories, take all appropriate measures to prevent the causing of significant harm to other watercourse States.

2. Where significant harm nevertheless is caused to another watercourse State, the States whose use causes such harm shall, in the absence of agreement to such use, take all appropriate measures, having due regard for the provisions of articles 5 and 6, in consultation with the affected State, to eliminate or mitigate such harm and, where appropriate, to discuss the question of compensation."

544 *Id.*, art. 12. Article 12, "Notification Concerning Planned Measures with Possible Adverse Effects," provides as follows: "Before a watercourse State implements or permits the implementation of planned measures which may have a significant adverse effect upon other watercourse States, it shall provide those States with timely notification thereof. Such notification shall be accompanied by available technical data and information, including the results of any environmental impact assessment, in order to enable the notified States to evaluate the possible effects of the planned measures."

545 *Wurttemberg and Prussia v. Baden* (*The Donauversinkug Case*), German Staatsgerichtshof, 1927, Entscheidungen des Reichesgereichts in Zivilsachen, v.116, appendix, 18–45, 86 Ann. Dig. 128 at 131.

Accordingly, in applying equitable and reasonable utilization, basin states must consider all relevant factors under specific circumstances including, *inter alia*, the cross-border effects of water use, hydrological conditions, populations dependent on the watercourse, states' social and economic needs, and environmental protection.[546] The factors in Article 6 of the UN Watercourses Convention are substantially similar to those in Article V[547] of the 1966 Helsinki Rules and Article VI[548] of the Water Charter. Article 7 reflects the obligation against causing transboundary harm derivative of customary international law.[549] The obligation of prior notification in Article 12 is framed in the 1992

546 UN Watercourses Convention, *supra* note 18, art. 6. The full text of Article 6, "Factors Relevant to Equitable and Reasonable Utilization" provides as follows:

"1. Utilization of an international watercourse in an equitable and reasonable manner within the meaning of article 5 requires taking into account all relevant factors and circumstances, including:

Geographic, hydrographic, hydrological, climatic, ecological and other factors of a natural character;

The social and economic needs of the watercourse States concerned;

The population dependent on the watercourse in each watercourse State;

The effects of the use or uses of the watercourses in one watercourse State on other watercourse States;

Existing and potential uses of the watercourse;

Conservation, protection, development and economy of use of the water resources of the watercourse and the costs of measures taken to that effect;

The availability of alternatives, of comparable value, to a particular planned or existing use.

2. In the application of article 5 or paragraph 1 of this article, watercourse States concerned shall, when the need arises, enter into consultations in a spirit of cooperation.

3. The weight to be given to each factor is to be determined by its importance in comparison with that of other relevant factors. In determining what is a reasonable and equitable use, all relevant factors are to be considered together and a conclusion reached on the basis of the whole."

547 1966 Helsinki Rules, *supra* note 260, art. V.

548 Article VI of the Water Charter is reproduced from Article V of the 1966 Helsinki Rules, mirrors the text and gives a list of the most important factors determining what is reasonable and equitable within the meaning of Article V of the Water Charter.

549 The *Trail Smelter* arbitration commented that: "[U]nder the principles of international law … no State has the right to use or permit the use of its territory in such a manner as to cause injury by fumes in or to the territory of another or the properties or persons therein, when the case is of serious consequence and the injury is established by clear and convincing evidence." *Trail Smelter* case (United States v. Canada) 3 UNRIAA 1911, 1938, 1963, 1966, 3 R. Int'l Arb. Awards 1905, 1965 (1941). Similarly, the ICJ in its oft cited *1996 Advisory Opinion on the Legality of the Threat or Use of Nuclear Weapons* recognized that "[t]he existence of the general obligation of States to ensure that activities within their jurisdiction and control respect the environment of other States or of areas beyond national

Rio Declaration on Environment and Development in the following terms: "States shall provide prior and timely notification and relevant information to potential affected States on activities that may have a significant adverse transboundary environmental effect and shall consult with those States at an early stage and in good faith."[550]

There are several other articles in the UN Watercourses Convention from which the indicia of a holistic approach to dealing with water resources can be gleaned. The UN Watercourses Convention's procedural obligations include cooperation,[551] data exchange,[552] and consultation (including EIAs and technical data).[553] Article 20 requires states to protect and preserve the

---

control is now part of the corpus of international law relating to the environment." 1996 ICJ 241, para. 29.

550 Rio Declaration, *supra* note 220, Principle 19.

551 UN Watercourses Convention, *supra* note 18, art. 8. Article 8, "General Obligation to Cooperate," provides:

"1. Watercourse States shall cooperate on the basis of sovereign equality, territorial integrity, mutual benefit and good faith in order to attain optimal utilization and adequate protection of an international watercourse.

2. In determining the manner of such cooperation, watercourse States may consider the establishment of joint mechanisms or commissions, as deemed necessary by them, to facilitate cooperation on relevant measures and procedures in the light of experience gained through cooperation in existing joint mechanisms and commissions in various regions."

552 *Id.*, art. 9. Article 9, "Regular Exchange of Data and Information," provides as follows:

"1. Pursuant to article 8, watercourse States shall on a regular basis exchange readily available data and information on the condition of the watercourse, in particular that of a hydrological, meteorological, hydrogeological and ecological nature and related to the water quality as well as related forecasts.

2. If a watercourse State is requested by another watercourse State to provide data or information that is not readily available, it shall employ its best efforts to comply with the request but may condition its compliance upon payment by the requesting State of the reasonable costs of collecting and, where appropriate, processing such data or information.

3. Watercourse States shall employ their best efforts to collect and, where appropriate, to process data and information in a manner which facilitates its utilization by the other watercourse States to which it is communicated."

553 *Id.*, art. 11. Article 11, "Information Concerning Planned Measures," states as follows: "Watercourse States shall exchange information and consult each other and, if necessary, negotiate on the possible effects of planned measures on the condition of an international watercourse."

"ecosystem"[554] of a watercourse. The goal of Article 21 is to reduce the amount of pollutants entering a watercourse.[555] For example, a tributary impacted by erosion may contribute significant quantities of silt and sediment to the mainstream thus destroying storage capacity and critical habitats.[556] Article 23 reflects the concept that watercourses do not function as isolated units.[557] For example, if a pollutant in a tributary makes its way to the mainstream that empties into a terminus such as a sea or gulf reentrant, the marine environment may become contaminated. Article 25 requires watercourse states to cooperate in the regulation of water flow in an international watercourse. This

554 *Id.*, art 20. Article 20, "Protection and Preservation of Ecosystems," provides: "Watercourse States shall, individually and, where appropriate, jointly, protect and preserve the ecosystems of international watercourses."

The ILC's commentary to article 20 indicates that an "ecosystem" is "an ecological unit consisting of living and non-living components that are interdependent and function as a community." *See* YBILC, vol.2, Part 2, at 118, para. 2 (1994).

555 *Id.*, art. 21. Article 21, Prevention, "Reduction and Control of Pollution," provides:

"1. For the purpose of this article, 'pollution of an international watercourse' means any detrimental alteration in the composition or quality of the waters of an international watercourse which results directly or indirectly from human conduct.

2. Watercourse States shall, individually and, where appropriate, jointly, prevent, reduce and control the pollution of an international watercourse that may cause significant harm to other watercourse States or to their environment, including harm to human health or safety, to the use of the waters for any beneficial purpose or to the living resources of the watercourse. Watercourse States shall take steps to harmonize their policies in this connection. Watercourse States shall, at the request of any of them, consult with a view to arriving at mutually agreeable measures and methods to prevent, reduce and control pollution of an international watercourse, such as:

(a) Setting joint water quality objectives and criteria;

(b) Establishing techniques and practices to address pollution from point and non-point sources;

(c) Establishing lists of substances the introduction of which into the waters of an international watercourse is to be prohibited, limited, investigated or monitored."

556 *See, e.g., Driscoll v. Adams*, 181 F. 3d 1285 at 1291 (11th Cir. 1999) wherein pollutant is broadly defined in the US Clean Water Act, 33 U.S.C. sec. 1251 *et seq.*, to include sand and silt.

557 UN Watercourses Convention, *supra* note 18, art. 23. The text of article 23, "Protection and Preservation of the Marine Environment," states as follows: "Watercourse States shall, individually and, where appropriate, in cooperation with other States, take all measures with respect to an international watercourse that are necessary to protect and preserve the marine environment, including estuaries, taking into account generally accepted international rules and standards."

provision has obvious implications for transboundary tributaries.[558] Finally, the UN Watercourses Convention provides a broad dispute settlement procedure to provide forward-thinking governments with constructive resolution of emerging environmental challenges at the intersection of law and policy, emphasizing strategic counseling, water diplomacy, crisis management, and creative dispute resolution.[559]

---

558 Under art. 25 of the UN Watercourses Convention, "Regulation":

"1. Watercourse States shall cooperate, where appropriate, to respond to needs or opportunities for regulation of the flow of the waters of an international watercourse.

2. Unless otherwise agreed, watercourse States shall participate on an equitable basis in the construction and maintenance or defrayal of the costs of such regulation works as they may have agreed to undertake.

3. For the purposes of this article, 'regulation' means the use of hydraulic works or any other continuing measure to alter, vary or otherwise control the flow of the waters of an international watercourse."

559 UN Watercourses Convention, art. 33. Article 33, "Settlement of disputes," provides:

"1. In the event of a dispute between two or more Parties concerning the interpretation or application of the present Convention, the Parties concerned shall, in the absence of an applicable agreement between them, seek a settlement of the dispute by peaceful means in accordance with the following provisions.

2. If the Parties concerned cannot reach agreement by negotiation requested by one of them, they may jointly seek the good offices of, or request mediation or conciliation by, a third party, or make use, as appropriate, of any joint watercourse institutions that may have been established by them or agree to submit the dispute to arbitration or to the International Court of Justice.

3. Subject to the operation of paragraph 10, if after six months from the time of the request for negotiations referred to in paragraph 2, the Parties concerned have not been able to settle their dispute through negotiation or any other means referred to in paragraph 2, the dispute shall be submitted, at the request of any of the parties to the dispute, to impartial fact-finding in accordance with paragraphs 4 to 9, unless the Parties otherwise agree.

A Fact-finding Commission shall be established, composed of one member nominated by each Party concerned and in addition a member not having the nationality of any of the Parties concerned chosen by the nominated members who shall serve as Chairman.

5. If the members nominated by the Parties are unable to agree on a Chairman within three months of the request for the establishment of the Commission, any Party concerned may request the Secretary-General of the United Nations to appoint the Chairman who shall not have the nationality of any of the parties to the dispute or of any riparian State of the watercourse concerned. If one of the Parties fails to nominate a member within three months of the initial request pursuant to paragraph 3, any other Party concerned may request the Secretary-General of the United Nations to appoint a person who shall not have the nationality of any of the parties to the dispute or of any riparian State

Collectively, these articles evidence a holistic or ecosystem approach in the UN Watercourses Convention, which McCaffrey says signals an "emerging principle of customary law in the field of shared freshwater resources: Namely, that states must protect the *ecosystems* of international watercourses."[560] In summary, the UN Watercourses Convention embodies a legal framework based on the concept of an international watercourse that is governed by the principle of equitable utilization, the no-harm rule, prior notification, and sustainable development.

In 2006, the World Wildlife Fund (WWF) undertook a campaign to establish the ILC Draft Articles on the Law of Transboundary Aquifers and Aquifer Systems as a protocol to the UN Watercourses Convention.[561] The Mekong countries were originally targeted in the WWF campaign. Unfortunately, due

---

of the watercourse concerned. The person so appointed shall constitute a single-member Commission.

6. The Commission shall determine its own procedure.

7. The Parties concerned have the obligation to provide the Commission with such information as it may require and, on request, to permit the Commission to have access to their respective territory and to inspect any facilities, plant, equipment, construction or natural feature relevant for the purpose of its inquiry.

8. The Commission shall adopt its report by a majority vote, unless it is a single-member Commission, and shall submit that report to the Parties concerned setting forth its findings and the reasons therefor and such recommendations as it deems appropriate for an equitable solution of the dispute, which the Parties concerned shall consider in good faith.

9. The expenses of the Commission shall be borne equally by the Parties concerned.

10. When ratifying, accepting, approving or acceding to the present Convention, or at any time thereafter, a Party which is not a regional economic integration organization may declare in a written instrument submitted to the Depositary that, in respect of any dispute not resolved in accordance with paragraph 2, it recognizes as compulsory ipso facto and without special agreement in relation to any Party accepting the same obligation:

(a) Submission of the dispute to the International Court of Justice; and/or (b) Arbitration by an arbitral tribunal established and operating, "unless the parties to the dispute otherwise agreed, in accordance with the procedure laid down in the annex to the present Convention. A Party which is a regional economic integration organization may make a declaration with like effect in relation to arbitration in accordance with subparagraph (b)."

560 Stephen C. McCaffrey, "An Overview of the U.N. Convention on the Law of the Non-Navigational Uses of International Watercourses" (2000) 20 *J. Land Resources and Envtl L.* 57, 70. (emphasis added).

561 Flavia Loures and Joseph Dellapenna, "Forthcoming Developments in International Groundwater Law: Proposals for the Way Ahead" (August 2007) 21 *Water Environment*

to budget constraints in 2009, the WWF initiative had to be refocused to concentrate on block ratifications in Europe and West Africa.[562] Under this new strategy in 2009, the Mekong countries were not prioritized at that time.[563] The UN Watercourses Convention required 35 ratifications or accessions before it entered into force and in 2011 there were only 24 contracting parties.[564]

When the UN Watercourses Convention was adopted by the GA, China, the regional hegemon and hydro-hegemon,[565] was one of only three governments that voted against it[566] (Figure 2.1)[567] citing the doctrine of absolute territorial sovereignty known as the Harmon Doctrine.[568] Noteworthy is that although all LMRB states voted in favor of the UN Watercourses Convention, only Vietnam has ratified it. Ratification of the 1997 UN Convention proceeded slowly. In April 2009, Tunisia became the 17th contracting state to the 1997 UN Convention.[569] In September 2009, Spain completed the ratification process.[570] In May 2010, Guinea-Bissau acceded to the treaty. On 27 September 2010, Nigeria ratified the Convention. On 2 December 2010, Greece became the 21st state to join the Convention. On 24 February 2011, France became the 22nd contracting state. Also in 2011, Burkina Faso and Morocco joined the Convention on 22 March and 13 April, respectively. From 30 April 2012 to 2 January 2015, the following states joined the UN Watercourses Convention: Montenegro, Denmark, Luxembourg, Benin, Chad, Italy, Niger, United Kingdom (UK), Ireland, Cote d'Ivoire, Vietnam, and State of Palestine. In August 2014, the UN Watercourses Convention entered into force with Vietnam as the 35th state joining the treaty. Irrespective

58, 58; Alistair Rieu-Clarke, "Entry into Force of the 1997 UN Watercourses Convention: Barriers, Benefits and Prospects" (December 2007) *Water21* 12.

562 *See* email from Flavia Loures, World Wildlife Fund [hereinafter WWF] International Water Law and Policy Senior Program Officer (email to author, April 4, 2009, 11:04 a.m. CST) (on file with author).

563 *Id.*

564 *Id.*; WWF, UN Freshwater Agreements Initiative, News Update, no. 07, at 3 (May 2009); WWF, UN Freshwater Agreements Initiative News Update, no. 13 at 2 (January 2011); WWF, UN Freshwater Agreements Initiative News Update, no. 14 at 2 (April 2011). The contracting states are Burkina Faso, Finland, France, Germany, Greece, Guinea-Bissau, Hungary, Iraq, Jordan, Lebanon, Libya, Morocco, Namibia, Netherlands, Nigeria, Norway, Portugal, Qatar, South Africa, Spain, Sweden, Syria, Tunisia and Uzbekistan.

565 *See generally,* Zeitoun and Warner, *supra* note 332.

566 Browder and Ortolano, *supra* note 241, at 526. Turkey and Burundi also voted against adoption of the 1997 UN Convention. Myanmar was absent for the vote.

567 Wolf, *supra* note 35, at 6.

568 UN General Assembly Press Release, GA/9248, 21 May 1997, at 5; *see also,* McCaffrey (1996), *supra* note 462, at 551.

569 WWF (May 2009), *supra* note 564, at 3.

570 WWF, UN Freshwater Agreements Initiative, News Update, no. 09 at 3 (November 2009).

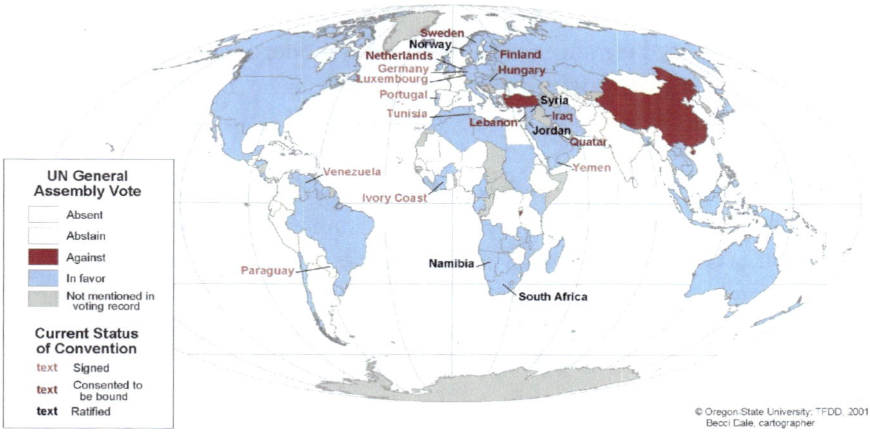

FIGURE 2.1 *On 21 May 1997, the United Nations General Assembly adopted the United Nations Convention on the Law of the Non-Navigational Uses of International Watercourses (UN Watercourses Convention) by 103 votes in favor, 3 against (China, Turkey and Burundi) and 27 abstentions.*
To bring the document into force, 35 instruments of ratification, acceptance, approval, or accession were necessary. Although all LMRB states voted for the 1997 UN Convention, only Vietnam has joined it. As of January 2015, there are 16 signatories and 36 parties to the UN Watercourses Convention. Theses signatories and parties are as follows: Benin, Burkina Faso, Chad, Cote d'Ivoire, Denmark, Finland, France, Germany, Greece, Guinea-Bissau, Hungary, Iraq, Ireland, Italy, Jordan, Lebanon, Libyan Arab Jamahiriya, Luxembourg, Montenegro, Morocco, Namibia, Netherlands, Niger, Nigeria, Norway, Paraguay, Portugal, Qatar, South Africa, Spain, State of Palestine, Sweden, Syrian Arab Republic, Tunisia, United Kingdom of Great Britain and Northern Ireland, Uzbekistan, Venezuela, Vietnam, and Yemen. The treaty entered into force in August 2014. Data sources: UN General Assembly Vote, United Nations (1997).
PRODUCT OF THE TRANSBOUNDARY FRESHWATER DISPUTE DATABASE (TFDD), COLLEGE OF EARTH, OCEAN, AND ATMOSPHERIC SCIENCES, OREGON STATE UNIVERSITY. MAP ORIGINALLY PUBLISHED IN *ATLAS OF INTERNATIONAL FRESHWATER AGREEMENTS* (UNEP, 2002). REPRINTED WITH PERMISSION FROM THE TFDD. ADDITIONAL INFORMATION ABOUT THE TFDD CAN BE FOUND AT: <HTTP://WWW.TRANSBOUNDARYWATERS.ORST.EDU>.

of its inordinately lengthy ratification process, the customary norms embodied in the UN Watercourses Convention were already being applied in multiple legal regimes. Indeed, the 1995 Mekong Agreement was drafted with the full benefit of the ILC draft articles of the UN Watercourses Convention, so analyzing the former through the lens of the latter is both a reasonable and logical approach.[571]

571 Stephen McCaffrey, "International Water Law for the 21st Century: The Contribution of the U.N. Convention" (January 2001) 118 *Water Resources Update* 11, 17.

2.2.4 *Waterscapes*

In the hydropolicy literature on the Mekong, a modern term, "waterscape,"[572] has emerged that is distinguished from a drainage basin, river basin and watercourse. A waterscape is:

> [T]he surface and groundwater resources of an area of land and their interrelationships with other physical, climatic and biotic elements, as well as with human activities. Waterscapes are an expression of the interaction between humans and their environment and encompass all of the social, economic and political processes through which water in nature is conceived of and manipulated by societies. In other words, waterscapes are landscapes viewed through the lens of their water resources, taken as a defining element of both ecosystems and human life.[573]

A waterscape is a landscape seen through its water attributes and its relationships with land, biota, and other elements. This approach facilitates the analysis of the transformations that occur in waterscapes from water diversions and interbasin transfers. Some commentators are beginning to question the efficacy of the river basin as the appropriate unit for water resources management.[574] That debate, although it underlines the social impact of tributary projects such as Yali Falls dam, is beyond the scope of this book but is an interesting one to monitor.

Waterscapes may provide a useful point of reference for legal analysis in the future.

2.2.5 *Salient Provisions of the 1995 Mekong Agreement*

Consisting of a preamble and 42 articles, the 1995 Mekong Agreement is contoured around the paradigm of sustainable development; recognizes the concept of sovereign equality and territorial integrity of the LMRB states; adopts the principle of equitable and reasonable use; provides for maintenance of mainstream flows; asserts the prevention and cessation of harmful effects; purports to provide for liabilities for damages; and contains limited procedures for dispute resolution. As previously mentioned, the treaty adopts a river

---

572 Francois Molle et al., "Introduction: Changing Waterscapes in the Mekong Region-Historical Background and Context" in Francois Molle et al. (eds.), *Contested Waterscapes in the Mekong Region: Hydropower, Livelihoods and Governance* (Earthscan, 2009) 1–19.

573 *Id.*

574 *See generally,* Jeroen Warner et al., "Going with the Flow: River Basins as the Natural Units for Water Management?" (2008) 10(Supp. 2) *Water Policy* 121.

basin approach which is interpreted to include the Mekong and its tributaries. Arguably, the Agreement also tacitly recognizes a community of interest of the LMRB states in the "Mekong River Basin" through its use of the phrase "community of Mekong nations."[575] The LMRB states agree:

> To cooperate in all fields of *sustainable development*, utilization, management and conservation of the water and related resources of the Mekong River Basin including, but not limited to irrigation, hydro-power, navigation, flood control, fisheries, timber floating, recreation and tourism, in a manner to optimize the multiple-use and mutual benefits of all riparians and to minimize the harmful effects that might result from natural occurrences and man-made activities.[576]

Thus, the modern Mekong legal regime goes well beyond promoting water resources development and control by going into the realm of water resources coordination for sustainable development.

Article 2 of the 1995 Mekong Agreement provides for the formulation of a Basin Development Plan (now Basin Development Strategy) but does not make it legally binding on the LMRB states:

> To promote, support, cooperate and coordinate in the development of the full potential of sustainable benefits to all riparian States and the prevention of wasteful use of Mekong River Basin waters, with emphasis and preference on joint and/or basin-wide development projects and basin programs through the formulation of a basin development plan, that would be used to identify, categorize and prioritize the projects and programs to seek assistance for and to implement at the basin level.

Whether IWRM, regional coordination, sustainable development, and holistic water resources management will be practically achieved under the Basin Development Strategy remains one of the most important questions in the LMRB. Arguably, IWRM cannot be accomplished unless tributaries are demarginalized

---

575 The Preamble to the 1995 Mekong Agreement states as follows in paragraph 6: "Affirming to promote or assist in the promotion of interdependent sub-regional growth and cooperation among the *community of Mekong nations*, taking into account the regional benefits that could be derived and/or detriments that could be avoided or mitigated from activities within the Mekong River Basin undertaken by this framework of cooperation." (emphasis added).

576 1995 Mekong Agreement, *supra* note 1, art. 1. (emphasis added).

in the Mekong legal regime and included in the political equation of basin-wide governance.

From an environmental perspective, the LMRB states pledge "[t]o protect the environment, natural resources, aquatic life and conditions, and ecological balance of the Mekong River Basin from pollution or other harmful effects resulting from any development plans and uses of water and related resources in the Basin."[577] Interestingly, water resources protection and use are explicitly subordinated to cooperation based on national sovereignty and territorial equality in the 1995 Mekong Agreement.[578] National sovereignty is the origin of the problem with marginalization of tributaries in the Mekong legal regime. Indeed, the challenge in negotiating the overarching provision of the 1995 Mekong Agreement, Article 5, was how to devise a water allocation system while not constraining national sovereignty.

Article 5 is the centerpiece of the 1995 Mekong Agreement under which the LMRB states agree:

> To utilize the waters of the Mekong River system in a reasonable and equitable manner in their respective territories, pursuant to all relevant factors and circumstances, the Rules for Water Utilization and Inter-basin Diversion provided for under Article 26 and the provisions of A and B below:
>
> A. *On tributaries of the Mekong River, including Tonle Sap, intra-basin uses and inter-basin diversions shall be subject to notification to the Joint Committee.*
>
> B. On the mainstream of the Mekong River:
>
> 1. During the wet season:
>   (a) Intra-basin use shall be subject to notification to the Joint Committee.
>   (b) Inter-basin diversion shall be subject to prior consultation which aims at arriving at an agreement by the Joint Committee.
> 2. During the dry season:
>   (a) Intra-basin use shall be subject to prior consultation which aims at arriving at an agreement by the Joint Committee.
>   (b) Any inter-basin diversion project shall be agreed upon by the Joint Committee through a specific agreement for each project prior to any proposed diversion. However, should there be a surplus quantity of water available in excess of the proposed uses of all

577 1995 Mekong Agreement, *supra* note 1, art. 3.

578 *Id.*, art. 4.

> parties in any dry season, verified and unanimously confirmed as such by the Joint Committee, an inter-basin diversion of the surplus could be made subject to prior consultation. (emphasis added).

This complex and convoluted allocation matrix was the result of intense negotiations among the LMRB states, particularly Thailand and Vietnam, and it is easy to see why Browder and Ortolano noted that Article 5 "was the most contentious issue during the negotiations and its complexity and ambiguity foreshadow the difficulties the Mekong [legal] regime may confront in dealing with water allocations."[579] The basic principle set out in Article 5 of the 1995 Mekong Agreement is equitable and reasonable utilization but this fundamental norm of international watercourse law is more adequately and fully set out in Article 5 of the UN Watercourses Convention where there is no differentiation between intrabasin and interbasin uses and diversions on the mainstream versus tributaries. Kaya has described the principle as follows:

> The principle of equitable utilization is grounded in the doctrine of limited territorial sovereignty and integrity within a given river basin. Under this principle, a basin state's sovereign rights to the waters of international rivers within or adjoining its territory are limited by the corresponding sovereign rights of other basin riparians. A state may thus utilize the water to the extent that this use does not interfere with the reasonable utilization of other basin states.[580]

Article 5(A), "the tributary rule,"[581] which controls all intrabasin uses and interbasin diversions of water on tributaries in the LMRB and requires only mere notice of such to the MRC JC before planned measures are undertaken, is derivative of a compromise solution between Thailand and Vietnam. This disparate treatment of tributaries, especially transboundary tributaries which are international rivers in their own right, is questionable in a river basin approach. Contributing significantly to this disparity in the Mekong legal regime is the

579 Browder and Ortolano, *supra* note 241, at 521.

580 Ibrahim Kaya, "The Euphrates-Tigris Basin: An Overview and Opportunities for Cooperation under International Law", (Arid Lands Newsletter no. 44, Fall/Winter 1998), available at <http://ag.arizona.edu/OALS/ALN/aln44/kaya.html> (accessed 15 June 2017); *See also* McCaffrey, *supra* note 9, at 384–405.

581 Article 5(A) of the 1995 Mekong Agreement is sometimes referred to hereinafter as "the tributary rule."

definition of the term "Proposed use," which in the 1995 Mekong Agreement excludes "domestic and minor uses of water not having a significant impact on mainstream flows."[582] Read in *pari materia* with Article 5(A), this definition, in effect, suggests that any "proposed uses" that affect only tributaries are excluded from notification procedures; in essence leaving tributaries wholly and fully outside the 1995 Mekong Agreement. The evolution of Article 5(A) is considered in Chapter 3 of this book.

In Article 5 of the 1995 Mekong Agreement, water uses in the LMRB are divided into two main categories, intrabasin uses and interbasin diversions.[583] Obligations of the LMRB states concerning notification, prior consultation, or agreement through the MRC JC in intrabasin uses or interbasin diversions from tributaries and the mainstream during the wet and dry seasons are covered in the PNPCA but are arguably not legally binding. Only in the case of an interbasin diversion from the mainstream in the dry season is an agreement by the LMRB states necessary.[584] "Notification" is defined in the 1995 Mekong Agreement as "[t]imely providing information by a riparian to the Joint Committee on its proposed use of water according to the format, content and procedures set forth in the Rules for Water Utilization and Inter-Basin Diversions under Article 26."[585] "Prior consultation" means:

> Timely notification plus additional data and information to the Joint Committee as provided in the Rules for Water Utilization and Inter-Basin Diversion under Article 26, that would allow the other member riparians to discuss and evaluate the impact of the proposed use upon their uses of water and any other affects, which is the basis for arriving at an agreement. Prior consultation is neither a right to veto the use nor unilateral right to use water by any riparian without taking into account other riparians' rights.[586]

---

582 1995 Mekong Agreement, *supra* note 1, Chapter II, para. 9. "Proposed use" is defined as "Any proposal for a definite use of the waters of the Mekong River system by any riparian, excluding domestic and minor uses of water not having a significant impact on mainstream flows."

583 Intrabasin uses are diversions of water within the LMRB, e.g., diverting water from the Sesan River at an upstream position and returning it at a downstream position. Interbasin diversions are transfers from the Mekong or a tributary to another basin.

584 1995 Mekong Agreement, *supra* note 1, art. 5(B)(2)(b).

585 *Id.*, Ch. II, para. 7.

586 *Id.*, Ch. II, para. 8.

The term "Agreement" used in Article 5 is defined in the Mekong legal regime as:

> A decision of the Joint Committee resulting from prior consultation and evaluation on any proposed use for inter-basin diversions during the wet season from the mainstream as well as for intra-basin use or inter-basin diversions of these waters during the dry season. The objective of this agreement is to achieve an optimum use and prevention of waste of the waters through a dynamic and practical consensus in conformity with the Rules for Water Utilization and Inter-Basin Diversions set forth in Article 26.[587]

The LMRB states assert their determination "[t]o utilize the waters of the Mekong River system in a reasonable and equitable manner in their respective territories, pursuant to [1] all relevant factors and circumstances, [2] the Rules for Water Utilization and Inter-basin Diversion provided for under Article 26 and [3] the provisions of [Article 5] A and B."[588] Article 6 of the 1995 Mekong agreement provides that the LMRB states agree to:

> [C]ooperate in the maintenance of the flows on the mainstream from diversions, storage releases, or other actions of a permanent nature; except in the cases of historically severe droughts and/or floods:
> A. Of not less than the acceptable minimum monthly natural flow during each month of the dry season;
> B. To enable the acceptable natural reverse flow of the Tonle Sap to take place during the wet season; and,
> C. To prevent average daily peak flows greater than what naturally occur on the average during the flood season.
>
> The Joint Committee shall adopt guidelines for the locations and levels of the flows, and monitor and take action necessary for their maintenance as provided in Article 26.

The Procedures for Maintenance of Flows on the Mainstream affirm that these flows will be maintained. Relative to Articles 5 and 6, Article 26 is mandatory and obligates the MRC JC to:

> [P]repare and propose for approval of the Council, inter alia, Rules for Water Utilization and Inter-Basin Diversions pursuant to Articles 5 and 6, including but not limited to; (1) establishing the time frame for the wet

587 *Id.*, Ch. II, para. 2.

588 *Id.*, art. 5.

and dry seasons; (2) establishing the location of hydrological stations, and determining and maintaining the flow level requirements at each station; (3) setting out criteria for determining surplus quantities of water during the dry season on the mainstream; (4) improving upon the mechanism to monitor intra-basin use; and, (5) setting up a mechanism to monitor inter-basin diversions from the mainstream.

Of particular relevance to development on tributaries are the PNPCA.[589] The definitions of Notification, Prior Consultation, and Agreement in these procedures mirror those in the 1995 Mekong Agreement. "Mekong Tributary," a term undefined in the Agreement text, in the PNPCA means:

> A natural stream of the Mekong River System. For the purposes of the present Procedures, a tributary as decided by the JC is a natural stream of the Mekong River System whose flows have a significant impact on the mainstream. This definition is subject to be reviewed and agreed upon after some time of implementation if any concern is raised.[590]

Notification to the MRC JC for intrabasin uses and interbasin diversions on tributaries is governed by Procedure 4 of the Procedures for Notification, Prior Consultation and Agreement.[591] Rule 5(A) of the 1995 Mekong Agreement

---

589 MRC, *supra* note 345, Procedures for Notification, Prior Consultation and Agreement, at 38–54.

590 *Id.* at 39, para. 2.

591 Procedure 4, "Notification," provides as follows:

"4.1 Scope of Notification

4.1.1 In accordance with Article 5 of the Mekong Agreement, notification on any proposed use stipulated in 4.1.2 shall be timely submitted to the MRC JC consistent with the format and content, schedules and principles prescribed in the Procedures, as appeared in Annex I.

4.1.2 Notification requirement and procedures shall be applied to the following proposed uses:

a. intra-basin use and inter-basin diversion on the tributaries, including Tonle Sap; and

b. intra-basin use during the wet season on the mainstream;

4.2 Content and Form/Format of Notification

4.2.1 Content

The Notification shall include feasibility study report, implementation plan, schedule and all available data.

4.2.2 Form/Format

To facilitate the notification formulation, the form/format for notification is provided as Annex I of the Procedures.

4.3 Institutional Mechanism for Notification

does not distinguish between water uses in the wet and dry seasons,[592] and requires mere notification to the JC irrespective of the type of use, whether

---

Mechanism for handling Notification under the Procedures shall involve National Mekong Committees (NMCs) and MRC's bodies with their respective roles/functions, responsibilities which are as follows:

4.3.1 The National Mekong Committee (NMC)

The roles/functions/responsibilities of each NMC under the Procedures are:

a. To inform the relevant line agencies of the scope, content and form for Notification of a proposed use as stipulated in 4.1 of the Procedures;

b. To review and check Notification received from line agencies concerned to ensure that data and information for Notification are complete and consistent with the content and form/format;

c. To assemble, record and transmit the Notification with appropriate documents to the MRC Secretariat for its submission to the MRC JC and transmission to the other NMCs.

4.3.2 The MRC Secretariat

The roles/functions/responsibilities of the MRC Secretariat under the Procedures are:

a. To receive, check for completeness, record and make files on the Notifications according to the form/format as appeared in Annex I;

b. To submit the Notification to the MRC JC and copy to each other NMCs;

c. To enter the relevant data and information into the MRC Secretariat Data and Information System; and

d. To place any comments on a Notification in the file, and submit to the MRC JC.

4.3.3 The MRC Joint Committee

The roles/functions/responsibilities of the MRC JC under the Procedures are to acknowledge any Notification submitted to it and take note of the comment, if any, submitted through the MRC Secretariat;

4.4 Process for Notification

Notification shall be transmitted by the relevant NMC to the MRC JC through the MRC Secretariat in conformity with their respective roles/functions/responsibilities as stipulated in 4.3 of the Procedures.

4.5 Timing for Notification

Notification of proposed use shall be transmitted to the MRC JC in a timely manner prior to implementation.

4.6 Absence of Notification

In case that the Notification has not been provided, the MRC JC will request the relevant NMC to fulfill its duties/responsibilities as provided in 4.3.1 of the Procedures."

592 Wet and dry seasons in the Procedures for Notification, Prior Consultation and Agreement are defined as: "The dates of the start and end of the wet and dry seasons vary throughout the basin due to regional variations. According to the preliminary analyses of the relatively long time series of hydrometeorological data, the wet season may start during mid-May to mid-June and end from mid-November to mid-December. The MRC JC will decide on the actual dates of the start and the end of the wet and dry seasons, based on analyses by the MRC Secretariat together with the National Mekong Committees (NMCs) of long term mainstream flow data."

an intrabasin use[593] or interbasin diversion. Procedure 4.2.1 implies that notification must be in writing because it "shall include [a] study report, implementation plan, schedule and all available data." Procedure 4.1.1 requires "[i]n accordance with Article 5 of the [1995] Mekong Agreement notification, on any proposed use stipulated in [procedure] 4.1.2 [to be] ... timely[594] [and to be] submitted to the MRC JC consistent with the format and content, schedules and principles prescribed in the Procedures ... in Annex I."[595] "Proposed

593 In the Procedures for Notification, Prior Consultation and Agreement, water use/utilization is defined as follows: "For the purpose of the present Procedures, it means any use of water which may have a significant impact to the water quality or flows regime of the mainstream of the Mekong River System by any member State. The Joint Committee may review and revise this definition from time-to-time as required for effective implementation of the Procedures."

594 Procedure 4.4, "Timing for Notification," of the Procedures for Notification, Prior Consultation and Agreement states: "Notification of proposed use shall be transmitted to the MRC JC in a timely manner prior to implementation."

595 The format of Annex I, "Form/Format of Notification," in the Procedures for Notification, Prior Consultation and Agreement is as follows:

"Annex I
Mekong River Commission Procedures for
Notification, Prior Consultation and Agreement
Form/Format of Notification
1. Notifying State
2. Date of submission:
3. Notifying Ministry/Agency (Name, address, telephone, fax,
e-mail):
4. Contact person/address (Name, address, telephone/fax/e-mail):
5. Name of the project:
6. Location of the project:
7. Nature of the proposed use:
a) *On tributary:*
- *Intra-basin uses*
- *Inter-basin diversion*
b) On the mainstream
- Intra-basin use during wet season
8. Purpose of the proposed project
9. Expected date for the implementation:
a) Date for starting the construction
b) Date for finishing the construction
c) Date for the operation
10. Duration and timing for water use
11. Description of the project (i.e.: scope, scale, site, type, quantity, capacity, and characteristic, etc.):
12. Attached documents": (emphasis added).

use" in the Procedures for Notification, Prior Consultation and Agreement is "[a]ny proposal for a definite use of the waters of the Mekong river system by any riparian, excluding domestic and minor uses of water not having a significant impact on mainstream flows."[596] The NMCs are responsible for transmitting notification to the MRC JC through the MRC Secretariat.[597] Since notification is the means via which the LMRB states inform each other of their intention to use water from a tributary, if notification is not provided, "the MRC JC will request the relevant NMC to fulfill its duties/responsibilities."[598] Seemingly then, by using the combined notification procedures in the PNPCA, each LMRB state may take as much water as it wants from its tributaries unless the mainstream is significantly impacted.

The 1995 Mekong Agreement does not contain detailed provisions on data and information exchange and sharing. The MRC JC is assigned the function "[t]o regularly obtain, update and exchange information and data necessary to implement this Agreement."[599] There is no legally binding provision in the 1995 Mekong Agreement or the Procedures for Data and Information Exchange and Sharing concerning information exchange on the possible effects of planned measures to the extent envisioned in the UN Watercourses Convention.[600]

Articles 6 and 26 of the 1995 Mekong Agreement require the LMRB states to maintain and manage the instream flows of the Mekong mainstream. Not surprisingly, these processes are also relegated to "procedures."[601] The Procedures for Maintenance of Flows on the Mainstream, as their title suggests, make no provisions for instream flow protection in tributaries.

Article 7 of the 1995 Mekong Agreement requires each LMRB state to avoid causing substantial damage to other riparians. It provides that the contracting parties are:

> To make every effort to avoid, minimize and mitigate harmful effects that might occur to the environment, especially the water quantity and quality, the aquatic (eco-system) conditions, and ecological balance of the river system, from the development and use of the Mekong River Basin

596 MRC, *supra* note 345, Procedures for Notification, Prior Consultation and Agreement at 40, Procedure 1, para. 10. This definition mirrors that in the 1995 Mekong Agreement. What constitutes a "significant impact on the mainstream" is undefined in the 1995 Mekong Agreement and the Procedures for Notification, Prior Consultation and Agreement.

597 MRC, *supra* note 345, Procedures for Notification, Prior Consultation and Agreement at 42, Procedure 4.4.

598 *Id.* at 42, Procedure 4.6.

599 1995 Mekong Agreement, *supra* note 1, art. 24(C).

600 1997 UN Convention, *supra* note 18, art. 11.

601 MRC, *supra* note 346, Procedures for Maintenance of Flows on the Mainstream.

> water resources or discharge of wastes and return flows. Where one or more States is notified with proper and valid evidence that it is causing substantial damage to one or more riparians from the use of and/or discharge to water of the Mekong River, that State or States shall cease immediately the alleged cause of harm until such cause of harm is determined in accordance with Article 8.[602]

Combined, Articles 3 and 7 provide the most comprehensive environmental protection ever adopted in the Mekong legal regime. Noteworthy is that Article 7 puts the burden on the victim state to prove with valid evidence that "substantial damage" is occurring.[603] The term "substantial damage" is not defined in the four corners of the 1995 Mekong Agreement but would seem to connote a higher level of "damage" than *significant harm* under the UN Watercourses Convention.

Articles 8,[604] 9,[605] and 10[606] of the 1995 Mekong Agreement generally comport with the principles of international watercourse law and international

---

602 1995 Mekong Agreement. *supra* note 1, art.7.

603 Compare art. 7, "Prevention and Cessation of Harmful Effects," in the 1995 Mekong Agreement, *supra* note 1 with art. 7, "Obligation Not to Cause Significant Harm," in the UN Watercourses Convention, *supra* note 18.

604 1995 Mekong Agreement, *supra* note 1, art. 8. Article 8, "State Responsibility for Damages," provides: "Where harmful effects cause substantial damage to one or more riparians from the use of and/or discharge to waters of the Mekong River by any riparian State, the party(ies) concerned shall determine all relative factors, the cause, extent of damage and responsibility for damages caused by that State in conformity with the principles of international law relating to state responsibility, and to address and resolve all issues, differences and disputes in an amicable and timely manner by peaceful means as provided in Articles 34 and 35 of this Agreement, and in conformity with the Charter of the United Nations."

605 *Id.,* art. 9. In Article 9, "Freedom of Navigation," the 1995 Mekong Agreement states:"On the basis of equality of right, freedom of navigation shall be accorded throughout the mainstream of the Mekong River without regard to the territorial boundaries, for transportation and communication to promote regional cooperation and to satisfactorily implement projects under this Agreement. The Mekong River shall be kept free from obstructions, measures, conduct and actions that might directly or indirectly impair navigability, interfere with this right or permanently make it more difficult. Navigational uses are not assured any priority over other uses, but will be incorporated into any mainstream project. Riparians may issue regulations for the portions of the Mekong River within their territories, particularly in sanitary, customs and immigration matters, police and general security."

606 *Id.*, art. 10. Article 10, "Emergency Situations," provides as follows: "Whenever a Party becomes aware of any special water quantity or quality problems constituting an emergency that requires an immediate response, it shall notify and consult directly with the

environmental law. Article 10 is noteworthy in that it seemingly demarginalizes tributaries to some degree by requiring an obligation to notify and consult on *any* emergency situations involving "special water quantity or quality problems." Although undefined in the 1995 Mekong Agreement, the UN Watercourses Convention sets out some of the latest thinking on exactly what an "emergency" is, defining it in Article 28 as "a situation that causes, or poses an imminent threat of causing, serious harm to watercourse States or other States and that results suddenly from natural causes, such as floods, the breaking up of ice, landslides or earthquakes, or from human conduct, such as industrial accidents." Articles 11–25 and Articles 27–33 cover the institutional framework of the MRC described in this book in Section 2.2.1.

Dispute resolution is addressed in the Agreement in Articles 34–35.[607] Article 34 of the 1995 Mekong Agreement specifies the MRC as the primary institutional mechanism for dispute resolution and requires the MRC to "make every effort to resolve the issue." If the MRC is unable to remedy the dispute, the next step is for "the Governments to take cognizance of the matter for resolution by negotiation through diplomatic channels within a timely manner."[608] Finally, Article 35 of the 1995 Mekong Agreement provides that "should the Governments find it necessary or beneficial ... they may ... request the assistance of mediation through an entity or party mutually agreed upon, and thereafter to proceed according to the principles of international law."

Noteworthy in Chapter VI, Final Provisions, of the Agreement are Articles 36–40. The 1995 Mekong Agreement "[r]eplace[s] the Statute of the

party(ies) concerned and the Joint Committee without delay in order to take appropriate remedial action."

607 *Id.*, arts. 34 and 35. Article 34, "Resolution by Mekong River Commission," provides: "Whenever any difference or dispute may arise between two or more parties to this Agreement regarding any matters covered by this Agreement and/or actions taken by the implementing organization through its various bodies, particularly as to the interpretations of the Agreement and the legal rights of the parties, the Commission shall first make every effort to resolve the issue as provided in Articles 18.C and 24.F."

Article 35, "Resolution by Governments," provides: "In the event the Commission is unable to resolve the difference or dispute within a timely manner, the issue shall be referred to the Governments to take cognizance of the matter for resolution by negotiation through diplomatic channels within a timely manner, and may communicate their decision to the Council for further proceedings as may be necessary to carry out such decision. Should the Governments find it necessary or beneficial to facilitate the resolution of the matter, they may, by mutual agreement, request the assistance of mediation through an entity or party mutually agreed upon, and thereafter to proceed according to the principles of international law."

608 *Id.*, art. 35.

Committee for Coordination of Investigations of the Lower Mekong Basin of 1957 as amended, the Joint Declaration of Principles for Utilization of the Waters of the Lower Mekong Basin of 1975, the Declaration Concerning the Interim Committee for Coordination of Investigations of the Lower Mekong Basin of 1978, and all Rules of Procedures adopted under such agreements."[609] The 1995 Mekong Agreement "may be amended, modified, superseded or terminated by the mutual agreement of all parties hereto at the time of such action."[610] Article 38 defines the Agreement's scope and leaves the possibility open for the LMRB states to enter into bilateral or multilateral agreements that do not conflict with it.[611] Article 39 provides for the future possibility of basin-wide water governance and integrated watershed management by leaving the 1995 Mekong Agreement open such that China and Myanmar by "accepting the rights and obligations under this Agreement, may become a party with the consent of the parties." Any LMRB state may "withdraw or suspend their participation under the present Agreement by giving written notice to the Chairman of the Council of the Mekong River Commission, who shall acknowledge receipt thereof and immediately communicate it to the Council representatives of all remaining parties."[612]

2.2.6 *The Sustainable Development Paradigm*

Holistic water resources management is a cornerstone of sustainable development. In the water resources spectrum, sustainable development is implemented through the principles of IWRM. The sustainable development paradigm is the presupposed nucleus of the Mekong legal regime.[613] Sustainable development is a concept articulated by the World Commission on Environment and Development in the 1987 Brundtland Report, defined as meeting the needs of the present without compromising the ability of future generations to meet their own needs.[614] It is at the confluence of three human

609 *Id.*, art. 36(B).

610 *Id.*, art. 37.

611 *Id.*, art. 38. Article 38, "Scope of Agreement," provides, *inter alia*, that: "[p]arties may enter into bi- or multi-lateral special agreements or arrangements for implementation and management of any programs and projects to be undertaken within the framework of this Agreement, which agreements shall not be in conflict with this Agreement and shall not confer any rights or obligations upon the parties not signatories thereto, except as otherwise conferred under this Agreement."

612 *Id.*, art. 40.

613 Pichyakorn, *supra* note 3, at 10.

614 *See* World Commission on Environment and Development, "Our Common Future" (Brundtland Report, 1987); Vick, *supra* note 25, at 225.

preoccupations: environment, economy, and society. Some commentators consider sustainable development to be a concept with no clear legal definition.[615] Others view sustainable development as a principle of international environmental law.[616] In the *Gabcikovo-Nagymaros* case, the ICJ invoked the concept of sustainable development in the context of the future legal regime of the Danube:

> Throughout the ages, mankind has, for economic and other reasons, constantly interfered with nature. In the past, this was often done without consideration of the effects upon the environment. Owing to new scientific insights and to a growing awareness of the risks for mankind—for present and future generations—of pursuit of such interventions at an unconsidered and unabated pace, new norms and standards have been developed, set forth in a great number of instruments during the last two decades. Such new norms have to be taken into consideration, and such new standards given proper weight, not only when States contemplate new activities but also when continuing with activities begun in the past. This need to reconcile economic development with protection of the environment is aptly expressed in the concept of sustainable development. For the purposes of the present case, this means that the Parties together should look afresh at the effects of the environment of the operation of the Gabcikovo power plant. In particular they must find a satisfactory solution for the volume of water to be released into the old bed of the Danube and into the side-arms on both sides of the river.[617]

In a separate opinion in the *Gabcikovo-Nagymaros* case, Vice President Weeramantry went further, stating that while the Court "referred to it as a concept ... I consider it to be more than a mere concept, but as a principle with normative value .... [T]he principle of sustainable development ..., in my view, ... is an integral part of modern international law."[618] "It is clearly of the utmost importance, both in case law and more generally."[619] According to Sands,

615 *See* U. Beyerlin, "The Concept of Sustainable Development" *in* R. Walfram (ed.), *Enforcing Environmental Standards: Economic Mechanisms as Viable Means?* (Springer, 1996) 95–121.

616 Philippe Sands, *Principles of International Environmental Law* (Cambridge University Press, 2nd ed., 2003) 253–254.

617 ICJ Reports 78 (1997), para. 140.

618 *Gabcikovo-Nagymaros* case, *supra* note 474. *See* Separate Opinion of Vice President Weeramantry, 1997 ICJ 7 at 82.

619 *Id.*

"[b]y invoking the concept of sustainable development, the ICJ indicates that the term has a legal function and both a procedural … aspect … and a substantive aspect."[620] Simply stated, "[i]nternational law recognizes a principle (or concept) of 'sustainable development'".[621] Sands points out though that "[t]he term needs to be taken in the context of its historic evolution as reflecting a range of procedural and substantive commitments and obligations."[622]

Birnie and Boyle describe seven substantive and procedural elements of sustainable development: (1) integration of environmental protection and economic development; (2) the right to development; (3) sustainable utilization and conservation of natural resources; (4) intergenerational equity;[623] (5) intragenerational equity;[624] (6) the polluter pays principle;[625] and (7) procedural elements of sustainable development—EIA, stakeholder participation, and access to information.[626] Applying these seven elements to international water law, Pichyakorn defines sustainable development in the international watercourse context as "the development of water in a manner in which [an] adequate supply of good quality water is sustained and the watercourse ecosystem is maintained for future generations."[627] She identifies five specific elements of "sustainable water development": (1) the right to use water; (2) the protection of water and prevention of water degradation; (3) the maintenance and conservation of water flow; (4) the ecosystem related approach; and (5) the procedural elements to achieve sustainable water development.[628] This element-driven approach allows resolution of current water use and quality problems and consideration of long-term environmental, social, and economic benefits concomitantly.

In the 1995 Mekong Agreement, sustainable development is interpreted as "a balance reached between the maintenance of the 'adequate quantity' and

---

620 Sands, *supra* note 616, at 255.

621 *Id* at 266.

622 *Id.*

623 *See* Edith Brown Weiss, "Our Rights and Obligations to Future Generations for the Environment" (1990) 84 *Am. J. Intl. L.* 198, 199. Intergenerational equity is the idea that as "members of the present generation, we hold the earth in trust for future generations."

624 Sands, *supra* note 616, at 253. Intragenerational equity is "the 'equitable' use of natural resources, which implies that use by one state must take account of the needs of other states."

625 *Id.* at 279. "The polluter-pays principle establishes the requirement that the costs of pollution should be borne by the person responsible for causing the pollution."

626 Birnie and Boyle, *supra* note 417 at 85–95; Pichyakorn, *supra* note 3, at 10.

627 Pichyakorn, *supra* note 3, at 10.

628 *Id.* at 11.

the preservation of 'good quality' of water."[629] To achieve this balance, articles on equitable utilization,[630] flow maintenance,[631] and environmental protection[632] are included in the Agreement. The 1995 Mekong Agreement also contains a provision requiring member states that cause substantial harm to other states to cease immediately the alleged activity causing the harm until the cause of the harm is determined.[633] These provisions, then, form the basic sustainable development balancing mechanism in the 1995 Mekong Agreement.[634]

The tendency of international watercourse agreements to increasingly adopt the sustainable development paradigm[635] does not guarantee that development will not trump environmental concerns. In light of the unbridled development on tributaries, the emerging issue in the LMRB is whether sustainable development in the 1995 Mekong Agreement is "a principle at odds with itself."[636] In the 1995 Mekong Agreement, the exact contours of sustainable development remain somewhat indeterminate. These observations are evidenced by the current practice in the LMRB. Indeed, one commentator has labeled the concept of sustainable development as "imported" into the Mekong regime by "both UN agencies and donor countries [who] have been instrumental in this programmatic shift" away from water allocation issues toward the emerging paradigm of sustainable development, which has "clashed with the existing decisional and development values ... of the Mekong riparian countries."[637]

2.2.7 *Balancing Water Quality and Water Quantity in the Mekong Regime*

The extent of success in achieving sustainable development in the LMRB is perhaps best viewed in the context of its core provisions: Articles 3, 5, 6, and 7. "Sustainable water development" in the Mekong regime means "the balance

629 Pichyakorn, *supra* note 43, at 175.

630 1995 Mekong Agreement, *supra* note 1, art. 5.

631 *Id.*, art. 6.

632 *Id.*, art. 3.

633 *Id.*, art. 7.

634 Pichyakorn, *supra* note 43, at 175.

635 *See, e.g.*, The 2002 Water Charter of the Senegal River, Charte des Eaux du Fleuvre Senegal, adopted by the Conference of Heads of State and Government, Organisation pour la Mise en Valeur Du Fleuve Senegal (OMVS), 18 May 2002, OMVS Resolution 005; Vick, *supra* note 25, at 225–227 and 233–237.

636 Stephen C. McCaffrey, "Introduction, Symposium: Taking Stock of Sustainable Development at 20: A Principle at Odds with Itself?" (2008) 21 *Pac. McGeorge Global Bus. and Dev. L.J.* 151.

637 Affeltranger, *supra* note 23, at 54.

reached between the adequate amount and good quality of the waters."[638] This means that the extent to which sustainable development can be achieved depends on "the effectiveness of putting Articles 5 and 6 (water quantity) and Articles 3 and 7 (water quality) into practice."[639] According to Pichyakorn, two situations may occur: (1) if all four articles are effectively implemented there is sufficient good quality water, or (2) if Articles 5 and 6 are effectively implemented but Articles 3 and 7 are not, while there will be sufficient water, the quality is degraded.[640] Pichyakorn explains how the consistent application of Articles 5 and 6 benefits the LMRB states:

> Articles 5 and 6 are quite rigid and likely to guarantee that the amount of water will be used equitably inasmuch as the flows are well maintained. The duty to notify, and arrange prior consultation ensure that the use of the quantity of the water will be monitored by all four countries. For instance, if the Member States would like to divert water … intra-basin in the dry season, they must arrange a prior consultation, which aims at arriving at an agreement by all members of the Joint Committee, or they are required to reach a specific agreement with the Joint Committee if they wish to conduct [an] inter-basin development project.[641]

Although water quality is inextricably linked to water quantity, the parties provided more flexibility in the former as evidenced by Article 7's relaxed standard allowing harmful effects but not "substantial damage," a term undefined in the 1995 Mekong Agreement. This raises the question whether Article 3 can achieve its purpose when the LMRB states are allowed to pollute water up to some undefined level. Also noteworthy is that Article 7 places the burden of proof upon the victim state.

In the framework of the 1995 Mekong Agreement, the parties were apparently more concerned with water quantity and allocation than water quality in framing the 1995 Mekong Agreement. This general observation follows from a comparison of Article 5 with Article 7. This relationship is mirrored somewhat by the disparity between Article 5 and Article 7 in the UN Watercourses Convention. Indeed, McCaffrey has described Article 7 as "the most controversial

638 Pichyakorn, *supra* note 43, at 176.

639 *Id.* at 176–177.

640 *Id.* at 177.

641 *Id.* (internal citations omitted).

provision of the entire Convention."[642] McCaffrey confirms that "the no-harm principle is not, and never has been, conceived as absolutely prohibiting the causing of significant harm under all circumstances."[643] Instead it "requires avoidance of harm in a way and to an extent that is reasonable under the circumstances."[644] Article 7 "plays a complementary role"[645] to Article 5 in the UN Watercourses Convention, and likewise in the 1995 Mekong Agreement. To some extent, then, equitable and reasonable utilization (water quantity) enjoys an elevated status in both agreements.

2.3 Chapter 2 Conclusion

In the 1950s, against a political backdrop of post-colonialism, insurgence, and wars of national liberation, the LMRB states, under the auspices of ECAFE (ESCAP), began exploring options for joint water resources development projects with the goal of erecting a cascade of large dams and reservoirs on the Mekong River. As a result, the four lower riparians concluded the 1957 Mekong Statute aimed at non-navigational and multi-purpose uses of this large river. The 1957 Mekong Statute gave the Mekong Committee the power to "promote, coordinate, supervise and control the planning and investigation of water resources development projects."[646] The principal objectives were to produce hydroelectric power, reduce flooding, alleviate poverty, increase dry season flows for navigation and irrigation, and procure international assistance and funding. Ironically, the only major hydropower dam constructed under the auspices of the initial Mekong Committee was not on the mainstream but rather a tributary, the Nam Ngum. China and Myanmar did not participate in this legal regime and this lamentably set the stage for piecemeal river basin management, which, for all practical purposes, continues today.

In 1971, work commenced on a Water Charter, which was the most comprehensive plan for international water law ever drafted in the Mekong legal regime. The Water Charter, which treated major tributaries like mainstream waters and provided for integration of minor tributaries into the mainstream hydraulic regime, evidenced significant political will by the LMRB states to use

642 Stephen C. McCaffrey and Mpazi Sinjelz, "The 1997 United Nations Convention on International Watercourses" (1998) 92 *Am. J. Intl. L.* 97, 100.

643 McCaffrey, *supra* note 9, at 407.

644 *Id.*

645 *Id.* at 408.

646 1957 Mekong Statue, *supra* note 244, art. 4.

a river basin approach for water resources utilization. A revised version of the Water Charter was later adopted as the 1975 Joint Declaration, which in retrospect was the most progressive agreement concerning Mekong tributaries ever signed. Considered *in pari materia* with the 1957 Mekong Statute, the 1975 Joint Declaration created an interdependency requiring unanimous approval by all basin states before implementation of all projects on major tributaries. The 1975 Joint Declaration provided that projects on major tributaries, like developments on the mainstream, must be approved by all LMRB states. Unfortunately, due to the onslaught of communism under the Pathet Lao movement in Laos, Viet Cong insurgency in Vietnam, and isolationist Khmer Rouge regime in Cambodia, the 1975 Joint Declaration was never put into practice. Not until 1978 would a groundswell of renewed cooperation, undoubtedly aimed at maintaining donor assistance and funding, reemerge in the LMRB when Laos, Thailand, and Vietnam concluded the Interim Mekong Committee Declaration.

The 1978 Interim Mekong Committee Declaration replaced the joint control of water resources in the LMRB with *promotion* of the development of water resources and is viewed largely as an interruption of the development of international water law in the LMRB. During the Interim Mekong Committee period, the Thai and Vietnamese began pursuing national agendas that proved contradictory and almost irreconcilable. Water use and allocation in the dry season emerged as the key issue in the complex and convoluted Mekong legal regime. Thailand insisted that the principle of unanimous approval be superseded allowing states to carry out national projects unilaterally on tributaries. Vietnam objected to any actions that would reduce dry season flow. China also entered the fray with their proposed eight dam Lancang cascade creating anxiety downstream. In the 1990s, Cambodia sought re-entry into the institutional framework. These tensions accentuated the need for a reformed, functional legal regime. The LMRB states needed a revolution in thinking—a fundamental re-evaluation of water allocation, law, and policy. A restructured water resources paradigm was imminent.

The 1995 Mekong Agreement was the first international water treaty to adopt the (then emerging) concept of sustainable development, the core of which purports to be harmonization of water quantity maintenance with water quality protection. As the outcome of continuous international cooperation spanning five decades, the 1995 Mekong Agreement and the MRC, the joint basin institution it sanctioned to manage the river, are often lauded as examples of best practices in international water law. However, more than 20 years of experience has shown that this vaunted treaty should be strengthened in order to live up to its name. The "imported" concept of sustainable development may be more related to placating donors than with functional hydropolicy. Using "basin" rhetoric, the treaty prioritizes sovereignty (Article 4), marginalizes tributaries (Article 5(A)), and provides for flow protection on the mainstream only (Article 6), vesting control of tributaries solely in the hands of the state-centric NMCs, thus allowing these vital arteries to be severed from their basin. The MRC has yet to develop detailed rules or procedures for water allocation and utilization or review of planned developments and flow maintenance on tributaries, preferring rather at every turn to redefine the basin, cling fervently to territorial integrity and sovereignty, broker megaprojects and attempt to harmonize dam construction and water diversions with "sustainability" concepts. These practices undermine ecosystem integrity, and when considered with a largely undefined legal mandate, evidence a marginalization of the MRC and call into question its legitimacy and relevance. Indeed on most tributary projects, the MRC continues to be sidestepped by the NMCs and now risks being superseded by the GMS programs.

Perhaps the most glaring deficiency in the Mekong legal regime is Article 5(A), "the tributary rule," that allows the LMRB states to use water in the tributaries however they choose, upon mere notice to the MRC JC, uncompromised by the substantive obligations in the 1995 Mekong Agreement. This article invites the question of whether the LMRB states—while pursuing hydrosovereignty on tributaries in a river basin *system* and supposedly those policies that would best ensure a multilateral legal regime's survival—can cope with the challenges and conflicts of uni- or bilateral development in the use of a common resource shared by states that are at different stages of political and economic development. Chapter 3 of this book traces the evolution of this regressive article and explores the disparate treatment of tributaries in the 1995 Mekong Agreement and other selected international and interstate legal regimes and case law.

CHAPTER 3

A Retrospective Look at Tributary Watercourse Law in the Mekong and Other Selected Legal Regimes

3.1 Evolution of Article 5(A) in the Mekong Legal Regime

By the 1990s, the shadow of transition was enveloping the LMRB, largely as the result of geopolitical shifts from post-Cold War relations. Cambodia had renewed interest in rejoining the Mekong regime. Nationalism began to churn the waters of Mekong tributaries. Water demand began to rival water supply in the LMRB. Water allocation became *the* emerging issue in Mekong hydropolicy, particularly in Thailand and Vietnam, shifting the focus away from the Mekong mainstream cascade, a project conceived in the 1950s and resurrected in 2006. Thailand became financially independent and no longer relied on donor funding for water projects. Accordingly, the Thai, with their surging economy, were less than enthused with a regime that required unanimous approval of national water projects by the three other LMRB states. The Indochina Federation, organized by Cambodia, Lao PDR, and Vietnam in 1975, continued to insist on joint approval for all Mekong water projects, a rule the Thai were determined to thwart.[647] Against this backdrop, water resources allocation in the Mekong legal regime underwent a significant paradigm shift.

Prior to the conclusion of the 1995 Mekong Agreement, development on the river's major tributaries was governed by Article 21 of the 1975 Joint Declaration, which provides that "A tributary recognized by all Basin States as a Major Tributary shall be considered as an integral part of the Basin development system and shall be governed by the provisions of the present Declaration of Principles applicable to the Mainstream." Mainstream waters are defined in Article 10 of the 1975 Joint Declaration as "a resource of common interest not subject to major unilateral appropriation by any riparian State without prior approval by the other Basin States through the Committee." Further, Article 20 requires agreement of all basin states for any extra-basin diversion of mainstream waters by a riparian state: "Extra-Basin diversion of mainstream waters by a riparian State shall require the agreement of *all* Basin States through a Project Agreement." (emphasis added). The original 1957 Mekong Committee regime included a "unanimity rule." Considered *in pari materia*, the 1957

647 Browder, *supra* note 250, at 66.

 | DOI 10.1163/9789004362604_005

framework and the 1975 Joint Declaration effectively gave the states veto power over planned measures on the mainstream and major tributaries. Article 5(A) of the 1995 Mekong Agreement provides that "[o]n tributaries of the Mekong River, including Tonle Sap, intra-basin uses and inter-basin diversions shall be subject [only] to notification to the [MRC] Joint Committee."[648] Consequently, compared with the 1975 Joint Declaration, the 1995 Mekong Agreement takes a step backwards, marginalizing tributaries. Development on Mekong tributaries then, is dealt with unilaterally or at best, bilaterally, rather than interdependently in the multilateral 1995 Mekong Agreement.

Article 5 of the 1995 Mekong Agreement, outlining the review protocol for proposed water uses, was the most difficult article to negotiate in the Mekong legal regime. This article emphatically shifted the focus away from the interdependent framework of the LMRB states under the 1957 Mekong Statute and 1975 Joint Declaration to national agendas and hydrosovereignty. The basic struggle was between the Thai, who were undertaking water diversion projects on the

648 1995 Mekong Agreement, *supra* note 1, art. 5. Article 5, "Reasonable and Equitable Utilization," in its entirety provides as follows: "To utilize the waters of the Mekong River system in a reasonable and equitable manner in their respective territories, pursuant to all relevant factors and circumstances, the Rules for Water Utilization and Inter-basin Diversion provided for under Article 26 and the provisions of A and B below:

A. On tributaries of the Mekong River, including Tonle Sap, intra-basin uses and inter-basin diversions shall be subject to notification to the Joint Committee. (emphasis added).

B. On the mainstream of the Mekong River:

1. During the wet season:
   (a) Intra-basin use shall be subject to notification to the Joint Committee.
   (b) Inter-basin diversion shall be subject to prior consultation which aims at arriving at an agreement by the Joint Committee.
2. During the dry season:
   (a) Intra-basin use shall be subject to prior consultation which aims at arriving at an agreement by the Joint Committee.
   (b) Any inter-basin diversion project shall be agreed upon by the Joint Committee through a specific agreement for each project prior to any proposed diversion. However, should there be a surplus quantity of water available in excess of the proposed uses of all parties in any dry season, verified and unanimously confirmed as such by the Joint Committee, an inter-basin diversion of the surplus could be made subject to prior consultation." Note the disparate treatment of tributaries in art. 5(A) requiring mere notification to the MRC JC for intrabasin uses and interbasin diversions compared with the requirements of prior consultation and agreement for certain Mekong mainstream water uses in art. 5(B). Because of the interconnectivity of tributaries with the mainstream, this distinction is obviously artificial.

TABLE 3.1 *General structure of Article 5(A) matrix*

Water use	Range of options for review of proposed water use
5(A) Tributaries	
1. Wet Season	
Intrabasin	Notification, Prior Consultation, or Agreement
Interbasin	Notification, Prior Consultation, or Agreement
2. Dry Season	
Intrabasin	Notification, Prior Consultation, or Agreement
Interbasin	Notification, Prior Consultation, or Agreement

MODIFIED FROM RADOSEVICH (1993A); RADOSEVICH AND OLSON (1999) AND BROWDER (1998).

Mekong, and the Vietnamese, who were insistent on maintaining dry season flows to prevent saltwater encroachment in the intensively farmed Mekong delta. The Thai rejected the veto power in the old regime, adopted a "no review requirements" approach and preferred that the customary international water law of prior notification, consultation, and avoidance of harm, embodied in the ILC's draft articles[649] finalized in 1994, be used in the new regime. In the Third Meeting of the MWG (MWG-III), UNDP legal advisor George Radosevich recorded the Thai stance: "Thailand is of the position that a country has the right to unilaterally develop its reasonable and equitable share of waters within its territory, based upon its own determination of that amount, but not to interfere with the reasonable and equitable share of the other riparians."[650] This position led to the development of a package agreement (matrix; Table 3.1) on Article 5(A)[651] instead of a choice "between the approach favored by the Thais–no review requirements (unless harmful effects are expected) or the approach favored by the Vietnamese–prior agreement for all projects."[652]

The Article 5(A) matrix (Table 3.1) divided water use on Mekong tributaries (separate and distinct from the mainstream) into seasonal (wet or dry) and location of use categories (intrabasin, within the Mekong Basin, or interbasin,

649 The ILC draft articles formed the basis of the negotiation of the UN Watercourses Convention, which hues closely to the draft articles.

650 George Radosevich, "Third Meeting of the Mekong Working Group (MWG-III): Report on Country Position Papers, Findings, Conclusion, Proposed Draft Agreement, and Meeting Results" (UNDP, Bangkok, Thailand, 7 July 1993) [hereinafter Radosevich (1993a)].

651 Browder, *supra* note 250, at 175.

652 *Id.* at 174.

outside the Mekong Basin), and three review procedures (notification, prior consultation, or agreement), which were not necessarily exclusive.[653] This disparate treatment of tributaries, the very arteries that supply half of the mainstream's flow, creates the unrealistic legal and hydrological fiction in the Mekong legal regime that the two are somehow not part of the same system, and thus, not interconnected. This reasoning is circular and fallacious. It also assumes there are no transboundary tributaries in the LMRB. Browder notes that:

> The distinction between the mainstream and the tributaries is based primarily on the issue of water rights. The Thai MWG felt that Thailand had the highest claim to tributaries of the Mekong River that originate in, and flow through, Thailand (no Thai tributaries flow through other basin states). The Thai MWG, however was willing to concede that all basin states had equal rights to water flowing in the mainstream ….
>
> From a hydrological perspective, the distinction between the tributaries and the mainstream is artificial because water in a tributary eventually becomes part of the mainstream. If each state had exclusive rights to its tributaries, then the potential exists for flows in the mainstream to be significantly reduced. In addition, there are several major tributaries that flow through two or more states in the Mekong Basin, including the Se Kong (Laos and Cambodia), Se San (Vietnam and Cambodia), and the Sre Pok (Vietnam and Cambodia).[654]

653 *Id.* at 175; 1995 Mekong Agreement, *supra* note 1, Ch. II-"Definition of Terms." para. 2, 7 and 8. The three procedural requirements of notification, prior consultation and agreement are defined in the 1995 Mekong Agreement as follows: "Notification: Timely providing information by a riparian to the Joint Committee on its proposed use of water according to the format, content and procedures set forth in the Rules" for Water Utilization and Inter-Basin Diversions under Article 26. Prior consultation: Timely notification plus additional data and information to the Joint Committee as provided in the Rules for Water Utilization and Inter-Basin Diversion under Article 26, that would allow the other member riparians to discuss and evaluate the impact of the proposed use upon their uses of water and any other affects, which is the basis for arriving at an agreement. Prior consultation is neither a right to veto the use nor unilateral right to use water by any riparian without taking into account other riparians" rights. Agreement under Article 5: A decision of the Joint Committee resulting from prior consultation and evaluation on any proposed use for inter-basin diversions during the wet season from the mainstream as well as for intra-basin use or inter-basin diversions of these waters during the dry season. The objective of this agreement is to achieve an optimum use and prevention of waste of the waters through a dynamic and practical consensus in conformity with the Rules for Water Utilization and Inter-Basin Diversions set forth in Article 26."

654 Browder, *supra* note 250, at 177.

Similarly, Carrard points out "[a]lthough [Article 5 of] the Agreement distinguishes between the Mekong River and its tributaries for management purposes, mechanisms for regulating tributary watercourse are sparse and lacking in clarity."[655] Commentary by Lebel and Garden confirms the convoluted and complex nature of this approach:

> The most difficult negotiation was on Article 5 on "*Reasonable and Equitable Utilization.*" Thai and the other teams had very different initial opinions; not surprisingly the final wording is complex, vague and inconsistent. A good example is the mainstream versus tributary distinction despite major tributaries in the basin flowing through multiple countries: a good example of politics of place and scale.[656]

Further, Lebel et al. (2008), posit that "[t]he basin as it is currently institutionalized under the Mekong Agreement is a superficial, truncated, and lopped vision of a 'basin'".[657] "Key provisions of the Mekong Agreement apply to the mainstem of the river and *not tributaries* and only to the four downstream riparian countries."[658]

Article 5(A) should be read and analyzed in the context of the entire 1995 Mekong Agreement, especially Articles 1 and 3 and the definition of "Proposed use" in Chapter II, "Definitions." Read in isolation, Article 1 refers to the "Mekong River Basin," an undefined term, and appears to promote a basin-wide scope encompassing the Mekong mainstream and its tributaries. Indeed, Rieu-Clarke and Gooch found that "[i]n terms of … [its] functional scope, the [1995] Mekong Agreement seeks to be all encompassing, thus covering 'all fields of sustainable development, utilization, management and conservation'".[659] They maintain that "it is therefore clear, at least from reading Article 1, that the 1995 Mekong Agreement covers both hydropower and its impact … [on Mekong tributaries]."[660]

---

655 Carrard, *supra* note 10.

656 Lebel and Garden, *supra* note 20, at 209–210 (internal citations omitted) (emphasis in original).

657 L. Lebel et al., "Beyond 'Basin': The Politics of Scale in Mekong Water Governance" (CGIAR Science Paper PA079, 2008), available at <https://www.researchgate.net/profile/Theib_Oweis/publication/266891806_A_methodology_for_the_assessment_of_agricultural_water_productivity_at_the_river_basin_level/links/543e4b670cf240f04d10f152.pdf#page=71> (accessed on 15 June 2017) [hereinafter Lebel et al. (2008)].

658 *Id.* (emphasis added).

659 Rieu-Clarke and Gooch, *supra* note 423, at 208.

660 *Id.* at 208–209.

Arguably, Article 3 of the 1995 Mekong Agreement, although not as detailed as its counterpart in the UN Watercourses Convention's Article 20, implies an ecosystem approach to protection of the Mekong from degradation as a result of any development plans or water uses in the *basin*, which again is interpreted to include tributaries. Noteworthy is McCaffrey's observation that "[i]t seems possible to conclude ... that there is now at least an emerging obligation to protect international watercourse systems and their *ecosystems* [including tributaries] against degradation."[661] He argues that "causing significant harm to the ecosystems of an international watercourse should be considered to be per se inequitable and unreasonable."[662]

A significant point of marginalization of Mekong tributaries in the 1995 Mekong Agreement is the stratagem employed in the definition of "Proposed use" in Chapter II. Article 5(A) obliges the LMRB States to notify the MRC JC of "intra-basin uses and inter-basin diversions" on tributaries. Cleverly, the treaty allows the parties to limit notification to "[a]ny proposal for a definite use of the waters of the Mekong River system ... excluding domestic and minor uses of water not having a significant impact on mainstream flows."[663] Arguably, a "Proposed use" on a tributary must have a significant impact on the flows of the Mekong proper before notification is required. In effect, this definition of "Proposed use" takes domestic and certain other projects on tributaries to the margins of the Mekong legal regime. Conveniently, the terms "domestic and minor uses" and "significant impact" are not defined. Key questions to consider then, are what the term minor use means and whether it limits the rest of the clause. For example, could a major dam be a "minor use" if it does not have a significant impact on mainstream flows. As previously mentioned in Section 2.2.5 of Chapter 2 of this book, it is Article 10 that effectively demarginalizes tributaries to some degree by bringing them back into the procedural (if not substantive) framework with mandatory notification and consultation only in the cases of an emergency.

Contrary to the old interdependent Mekong legal regime (17 September 1957–4 April 1995), then, the LMRB states proceeded into their negotiations about the Article 5 matrix under the general principle "that water in the Mekong River is a common resource of all basin states, whereas states have an (undefined) higher priority for the use of water in tributaries within their territory."[664] Negotiations on the Article 5 matrix were protracted, lasting more

661 McCaffrey, *supra* note 9, at 457.

662 *Id.* at 458.

663 1995 Mekong Agreement, *supra* note 1, Chapter II, para. 9.

664 Browder, *supra* note 250, at 180.

than a year from August 1993 to November 1994.[665] Article 5 was the subject of multiple revisions of a text drafted by UNDP legal advisor George Radosevich, whose notes provide critical insight into the evolution of the tributary rule.[666]

At TDM-I in August 1993, the Thais proposed an original definition of notification, the current procedural requisite for any planned measure on tributaries in the current Mekong legal regime, as follows: "Notification: refers to the format and content adopted by the [Joint] Committee for a riparian to inform the [Joint] Committee of uses, *without any discussion necessary*, assuming the notification is timely and complete according to the Committee's Rules."[667] Because of the concern over decreased flows, the Vietnamese MWG was not comfortable with this definition.[668] Consequently, the term "without any discussion necessary" was removed, and the 1995 Mekong Agreement defines "notification" as "[t]imely providing of information by a riparian to the Joint Committee on its *proposed use* of water according to the format, content and procedures set forth in the Rules for Water Utilization and Inter-Basin Diversions under Article 26."[669]

The Cambodians entered into the fray at TDM-I and insisted that the Tonle Sap be specifically included in the tributary rule.[670] The Tonle Sap, which reverses course with the annual wet season flows, is dependent on those flows from the Mekong mainstream to replenish its waters. But since the mainstream is considered a common resource of all LMRB states, classifying the Tonle Sap as a tributary within Cambodian territory provided greater priority of use and control over its water resources. Thailand and the Lao PDR agreed to this classification, but Vietnam, in the spirit of continuing negotiations, only reluctantly acceded because of the its own concern about vital dry season flows from the lake into the Mekong delta.[671]

Prior to the MWG-IV meeting, the national positions on review procedures for water uses and diversions on tributaries in Article 5(A) were (1) notification for all uses and diversions in the wet season; and (2) in the dry season,

---

665 *Id.*

666 George Radosevich, "Draft Agreement from Technical Drafting Meeting (TDM)" (UNDP, Bangkok, Thailand, 4 September 1993) [hereinafter Radosevich (1993b)].

667 *Id.* at 19 (emphasis added).

668 Browder, *supra* note 250, at 182.

669 *Id.*; 1995 Mekong Agreement, *supra* note 1, Chapter II-Definition of Terms, para. 7. (emphasis in original).

670 Radosevich (1993b), *supra* note 666, at 6.

671 Browder, *supra* note 250, at 180–181.

TABLE 3.2 *Lower Mekong River Basin states' positions on Article 5(A) prior to Mekong Working Group-IV*

Water use	Position on review procedure
5(A) Tributaries	
1. Wet Season	Notification
2. Dry Season	
Intrabasin	Notification
Interbasin	Prior Consultation

MODIFIED FROM RADOSEVICH (1993C) AND BROWDER (1998).

TABLE 3.3 *Vietnam's proposal on Article 5(A) at Technical Drafting Meeting-II*

Water use	Proposed review procedure
5(A) Tributaries	
1. Wet Season	Notification (Agreement on all interbasin diversions)
2. Dry Season	Prior Consultation (Agreement on all interbasin diversions)

MODIFIED FROM RADOSEVICH (1994) AND BROWDER (1998).

notification for intrabasin uses and prior consultation for interbasin diversions (Table 3.2).[672]

Later, at TDM-II, the Vietnamese proposed a new position on review procedures on tributaries, one requiring an agreement on *all* interbasin diversions (Table 3.3).[673] This new position caused an impasse in negotiations between Thailand and Vietnam.[674] After an inordinately lengthy delay in negotiations, the Thai Prime Minister visited and consulted with his counterpart in Vietnam.[675] The Thai position on the deadlock was captured in a magazine interview later:

672 *Id.* at 183; George Radosevich, "MWG-IV Report and Request for Action" (UNDP, Bangkok, Thailand, 12 December 1993) 1 [Radosevich (1993c)].

673 George Radosevich, "Minutes of the TDM-II" (UNDP, Bangkok, Thailand, 6–7 January 1994); Browder, *supra* note 250, at 185.

674 Browder, *supra* note 250, at 186.

675 *Id.* at 187.

> [The Thai government] is of the opinion that it should be able to utilize water from tributaries without having to consult with any international organization. If Thailand wants to build a dam across the Kok River [for the Mekong-Chao Phraya scheme] and has to consult the committee, it is not right .... Vietnam has the responsibility to ask for another meeting to resolve the impasse, but they have not done so yet. If there is going to be a meeting, we are very happy to join in. But the clear standing point for Thailand is that we will continue from what we have already agreed upon. We are not going to waste time to reconsider the principles we have already agreed upon.[676]

In May 1994, the Indochina Federation met in Ho Chi Minh City and reformulated Article 5(A) (Table 3.4) to require mere notification prior to all water uses and diversions on Mekong tributaries, distinguishing it from the TDM-II proposal requiring prior consultation for intrabasin dry season use.[677]

In a letter from the Vietnamese MWG to the UNDP, which was transmitted to the Thais, the former country proposed that: "Riparian states are entitled to use of water within their respective territories. However, since the use of waters in tributaries have [sic] impact on the mainstream waters, states need to notify one another of their plans of water use in tributaries."[678] The May 1994 proposal was "warmly received by the Thai MWG and eventually served as the basis for the final [rectified] version of Article 5[A] in the Mekong [legal regime]."[679] Weatherbee concisely summarizes the obvious effects of the final Article 5 (A):

TABLE 3.4 *Indochina Federation (Cambodia, Lao PDR, and Vietnam) May 1994 Joint Proposal on Article 5(A)*

Water use	Proposed review procedure
5(A) Tributaries	Notification

MODIFIED FROM VIETNAM NMC (1994) AND BROWDER (1998).

676 *Id.* at 187–188 (citing "Thai and Vietnam Disagree to Investigate the Lower Mekong Basin" translated from the Thai periodical *Manager Magazine* (28 March 1994).

677 *Id.* at 188; See Vietnam National Mekong Committee, Letter to UNDP Resident Representatives in Vietnam, UNDP, Bangkok, Thailand (12 May 1994).

678 *Id.* at 189 (paraphrased from the Vietnam National Mekong Committee, Letter to UNDP Resident Representative in Vietnam, UNDP, Bangkok, Thailand (12 May 1994)).

679 *Id.* at 189.

> The tension between upstream and downstream national interests is the most politically sensitive obstacle to cooperation.
>
> The terms of reference to the upstream-downstream riparian interdependencies have changed in the new Mekong Committee regime. The 1975 principles for the utilization of the Mekong held that projects involving major tributaries of the basin as well as extra-basin [interbasin] diversion of mainstream flow must be approved by all riparian states. At the insistence of Thailand, this principle has been superseded, with individual countries having the right to carry out national projects unilaterally without the approval of the other riparian states .... The Thai position is uncompromising. Bangkok's lever ... was the threat of nonparticipation.[680]

It took 21 months to negotiate and complete the 1995 Mekong Agreement. Over one year of that time was focused on Article 5, which dominated the negotiations. In the end, the Thais succeeded in removing the veto power and softening the Mekong legal regime, which today stands as only a surviving trace or vestige of the earlier 1957 Mekong Statute, Water Charter, and the 1975 Joint Declaration. Article 5(A)'s marginalization of tributaries, in effect, thwarts holistic water resources management, basin-wide governance, and sustainable water development, and is transforming the LMRB into a waterscape sliced by dams and electricity infrastructure.

3.2 Tributary Watercourse Law in Other Selected Legal Regimes

3.2.1 *The 1909 Boundary Waters Treaty (US and Canada)*

Viewed in the context of territorial integrity and national sovereignty, the marginalization of tributaries in the Mekong legal regime may appear to be the acceptable norm. Indeed, several international watercourse agreements and interstate acts contain examples of disparate treatment of tributaries. For example, McCaffrey notes this disparity in his analysis of the 1909 Boundary Waters Treaty which narrowly defines boundary waters as:

> [T]he waters from main shore to main shore of the lakes and rivers and connecting waterways, or the portions thereof, along which the international boundary between the United States and the Dominion of Canada passes, including all bays, arms, and inlets thereof, *but not including tributary waters* which in their natural channels would flow into such lakes,

680 Weatherbee, *supra* note 46, at 175.

> rivers, and waterways, or waters flowing from such lakes, rivers, and waterways, or the waters of rivers flowing across the boundary.[681]

The 1909 Boundary Waters Treaty creates a reservation in favor of each state (US and UK, acting for Canada at the time) for jurisdictional control of tributaries while providing an equality of rights to boundary waters.[682] This disparate treatment of tributaries seemingly evidences a preference for at least a partial incorporation of the Harmon Doctrine, which is tantamount to a right to act unilaterally on tributaries in pursuit of national sovereignty. Noteworthy in the negotiations was that the US maintained that tributaries should not be included in the International Joint Commission's authority,[683] while the UK (Canada) preferred a comprehensive agreement which would include them.[684] However, McCaffrey suggests that this view is "misplaced" and is merely unilateralism on paper as the US had repudiated the Harmon doctrine.[685] He cites the subsequent 1961 Columbia River Treaty as an example of the two states' ability to equitably apportion the benefits of that river.[686] As McCaffrey aptly points out, "a state's sovereignty over its rivers does not give it license to harm other

681 McCaffrey, *supra* note 9, at 347; 1909 Boundary Waters Treaty, *supra* note 13, preliminary art. (emphasis added).

682 A. Dan Tarlock, "International Water Allocation", sec. 11:14 in *Law of Water Rights and Resources* (Thomson Reuters, 2001).

683 The International Joint Commission governs the relations of the US and Canada under the 1909 Boundary Waters Treaty. The International Joint Commission has authority over boundary waters while the US and Canada have control over tributaries. It has two broad powers-judicial and investigative. The International Joint Commission's approval is required for hydropower projects and water diversions on boundary waters that will "raise the natural level of waters on the other side of the boundary." *See* 1909 Boundary Waters Treaty, *supra* note 13, art. IV. The International Joint Commission has the power to recommend a resolution to any binational disputes over boundary waters.

684 Richard Paisley, "Adversaries Into Partners: International Water Law and the Equitable Sharing of Downstream Benefits" (2002) 3(2) *Melbourne J. Int. Law* 280, 285.

685 McCaffrey, *supra* note 9, at 107–108. The US State Department has expressly conceded the US considered the Harmon Doctrine to be a special pleading and repudiated it. *See* "Memorandum to the Legal Advisor" (November 23, 1942) reprinted in Marjorie M. Whiteman, *Digest of International Law* (U.S. Department of State, vol. 3 1964) 950. *See also* U.S. Department of State, "Legal Aspects of the Use of Systems of International Waters," *in* Sen. Doc. No.118, 85th Cong., 2nd Sess. 43 (1958) 89–91; *See generally,* David R. Deener, *The United States Attorney Generals and International Law* (The Hague, Martinus Nijhoff, 1957) 253–257 and 308–309.

686 McCaffrey, *supra* note 9, at 351–353.

states."[687] This point was recognized by Chandler P. Anderson, an American negotiator of the 1909 Boundary Waters Treaty, who in a communiqué to the Secretary of State stated:

> [A]bsolute sovereignty carries with it the right of inviolability as to such territorial waters, and inviolability on each side imposes a coextensive restraint upon the other, so that neither country is at liberty to so use its own waters as to injuriously affect the other.
>
> [T]he conclusion is justified that international law would recognize the right of either side to make any use of the waters on its side which did not interfere with the coextensive rights of the other, and was not injurious to it.[688]

Arguably, then, state A should not cause injury to state B by means of planned measures on tributaries wholly within the former's territory. This can best be ensured by imposing some limitation or constraint on sovereignty. The 1909 Boundary Waters treaty does just that in Article II by significantly qualifying unilateral actions.

In comparison and contrast with the 1995 Mekong Agreement, unilateral action in the 1909 Boundary Waters treaty is actually somewhat limited or qualified because "any interference with or diversion ... of such waters ... resulting in any injury on the other side of the boundary, shall give rise to the same rights and entitle the injured parties to the same legal remedies as if such injury took place in the country where such diversion or interference occurs."[689] Paisley

687 *Id.* at 103.

688 Quoted in "Memorandum in Relation to the Arbitration of the *Trail Smelter* Case, United States and Canada", August 10, 1937 (prepared by Green H. Hackworth, Legal Advisor, for Swagar Sherley, Agent of the United State) in Whiteman, *Digest of International Law*, (U.S. Department of State, vol. 5, 1963–1970) 60–61; McCaffrey, *supra* note 9, at 582. (citation omitted).

689 1909 Boundary Waters Treaty, *supra* note 13, art. II. Article II declares "[e]ach of the High Contracting Parties reserves to itself or to the several State Governments on the one side and the Dominion or Provincial Governments on the other as the case may be, subject to any treaty provisions now existing with respect thereto, the exclusive jurisdiction and control over the use and diversion, whether temporary or permanent, of all waters on its own side of the line which in their natural channels would flow across the boundary or into boundary waters; but it is agreed that any interference with or diversion from their natural channel of such waters on either side of the boundary, resulting in any injury on the other side of the boundary, shall give rise to the same rights and entitle the injured parties to the same legal remedies as if such injury took place in the country where such

describes the political dynamics in the negotiations as "reflect[ing] to some extent the interests of each negotiating state."[690] He observes that:

> The issue of most concern to the US was sovereignty. The US, while realising the necessity of an agreement to manage transboundary waters, wanted to ensure that its political independence was not compromise[d] in the process. This was expressed in the US position that absolute territorial sovereignty must be retained over the waters within each state's territory. It was the view of the US that tributaries should not be included in the new commission's authority. In contrast, Canada was interested in establishing an egalitarian relationship with the US. Canada was hampered in its pursuit not only by the relative size and level of development of the two states at the time, but also because Canadian foreign policy was still the purview of the United Kingdom. Consequently, negotiations had to be carried out between Ottawa, Washington and London. Generally, however, Canada wanted a comprehensive agreement, which would include tributaries, and a commission with greater authority than former bodies.[691]

Thus, the exclusion of tributaries in the 1909 Boundary Waters Treaty predicated on US national sovereignty is somewhat dissimilar from the marginalization of tributaries in the 1995 Mekong Agreement, because the former agreement excludes tributaries but is tempered by a specific qualification for *any* unilateral injurious interference or diversion.[692] Article II of the 1909 Boundary Waters

---

diversion or interference occurs; but this provision shall not apply to cases already existing or to cases expressly covered by special agreement between the parties hereto. It is understood however, that neither of the High Contracting Parties intends by the foregoing provision to surrender any right, which it may have, to object to any interference with or diversions of waters on the other side of the boundary the effect of which would be productive of material injury to the navigation interests on its own side of the boundary," albeit it should be noted that the rights and remedies referred to in Article II are those of private parties to be vindicated through the courts and presumably not those of the parties to the treaty.

690 Paisley, *supra* note 684, at 285 (citing Aaron Wolf, "Transboundary Waters: Sharing Benefits, Lessons Learned" (Draft Thematic Background Paper, International Conference on Freshwater, 2001) available at <http://lv-twk.oekosys.tu-berlin.de/project/inwent/images/pdfs/transboundary-waters.pdf> (accessed 15 June 2017).

691 *Id.* (internal citations omitted).

692 1909 Boundary Waters Treaty, *supra* note 13 art. II; 1995 Mekong Agreement, *supra* note 1, arts. 5(A) and 7.

Treaty, in essence, affirms the interdependence of all waters in the Canadian and US borders water resources spectrum. This then, constitutes a qualified exclusion of tributaries as contrasted with marginalization. Like the 1995 Mekong Agreement, the 1909 Boundary Waters Treaty obfuscates the concept of unity of a drainage basin by separating boundary waters from tributaries. However, in the final analysis, the 1909 Boundary Waters Treaty preserves some degree of interdependence of all basin waters in Article II by placing a significant limitation on unilateral action on tributaries which might cause transboundary harm, and practice under the 1909 Treaty has effectively changed the definition to include tributaries.[693] The same is not true in the Mekong legal regime.

3.2.2 *The 1961 Columbia River Treaty (US and Canada) and the Okanagan River Basin*

A prime example of the disparate treatment of a tributary watercourse in the US-Canadian boundary waters spectrum is that of the Okanagan River in the bilateral 1961 Columbia River Treaty. The Columbia, one of the largest rivers in the world,[694] rises in British Columbia, flows through that country and then south into the US in Washington State before forming part of the border between Washington and Oregon.[695] There are three main transboundary rivers in the Columbia watershed: from east to west, the Kootenay (Kootenai in the US), the Columbia, and the Okanagan. The Kootenay, Okanagan, Similkameen, and Kettle rivers are the major Canadian tributaries to the Columbia. The Okanagan valley is an area of burgeoning population growth, in stark contrast to the empty wilderness just downstream of the US-Canadian border.

The Okanagan River rises in southern British Columbia and drains a plateau region east of the Cascade Range and north and west of the Columbia and the Okanagan Valley. The Okanagan River enters the Columbia River from the north, 8 km east of Brewster, Washington, between Wells dam (downstream) and Chief Joseph dam (upstream). The Okanagan empties into Lake Pateros,

693 *See, e.g.*, "Impacts of a Proposed Coal Mine in the Flathead River Basin" (1983).

694 Gupta, *supra* note 31, at 3.

695 John V. Krutilla, *The Columbia River Treaty: The Economics of an International River Basin Development* (Johns Hopkins Press, 1967) 2–5; Keith W. Muckleston, "International Management in the Columbia River System" (Technical Document in Hydrology, UNESCO SC-2003/WS/56, 2003) 2–8.

the reservoir behind Wells dam. The river is 314 km long and drains an area of 21,600 km^2.[696]

Derivative of the dispute surrounding construction of the Libby dam on the Kootenay tributary, the 1961 Columbia River Treaty between the US and Canada was signed in 1961 and entered into force in 1964. The Treaty has been lauded as a model of 45 years of US-Canadian continuing cooperation, especially for its principle of equality and equitable apportionment of benefits.[697] The 1961 Columbia River Treaty is highly technical and is concerned primarily with two purposes: hydropower development and flood control. The Columbia River is the most hydroelectrically developed river system in the world with 219 major dams located on its tributaries.[698] The Treaty deals with power production and floods on the Kootenay-Columbia watercourse. Although lauded for its principle of equality and equitable apportionment of benefits, in reality, to the detriment of basin-wide governance, the treaty does not deal with the other tributaries of the Columbia and indeed excludes one of its main tributaries, the Okanagan River. The Treaty does not mention the Okanagan largely because it is the driest watershed in Canada and was left out apparently because it likely did not have enough water for hydropower production compared to the Columbia's other tributaries.[699] This is an unqualified exclusion of a tributary watercourse. In contrast to the apparent neglect of the Okanagan basin in the 1961 Columbia River Treaty, there was at least considerable attention given to Mekong tributaries in the debates surrounding the negotiations for the 1995 Mekong Agreement. While true that the Mekong legal regime does not explicitly address transboundary tributaries, it was not for lack of discussion. The Water Charter, 1975 Joint Declaration, and draft documents in the negotiations of the 1995 Mekong Agreement were far more inclusive of the unitary basin concept than Rule 5(A) as ultimately adopted in the current Mekong legal regime. However, charitably read between its lines, the 1995 Mekong Agreement is broad enough to include transboundary tributaries by enabling the LMRB states and MRC to adopt rules, procedures, and guidelines concerning these important river arteries. The same cannot be said about the 1961 Columbia

696 Summary of US-Canadian Water Agreements *in* J.P. Bruce et al., "Climate Change Impacts on Boundary and Transboundary Water Management" (Climate Change Action Fund Project Natural Resources Canada Project No. A458/402, 30 June 2003) 2.

697 McCaffrey, *supra* note 9 at 353.

698 Rachael Paschal Osborn, "Columbia River Treaty: Past and Possible Futures" (WA-AWRA Section Newsletter, November-December 2007) 9.

699 Hans Schrier, University of British Columbia Institute for Resources, Environment and Sustainability (email communiqué to author, November 6, 2007, 11:27 a.m. CST) (on file with author).

River Treaty, in which the entire Okanagan River was simply lopped off and never mentioned. Although it purports to harmonize the river's legal regime, the 1961 Columbia River Treaty fails to consider one of its major tributaries and therefore does not embrace the international drainage basin as a unitary whole.

Hydropower development is accelerating on tributaries throughout the LMRB. Accordingly, and perhaps ironically, from 28 April through 2 May 2008, the MRC sent a delegation on a "high level study visit" to the Columbia River basin.[700] This study had the following three objectives:

(1) To learn from the positive and negative economic, social, environmental, political and cultural effects of extensive and large-scale water resources development in order to enhance the shared vision of sustainable water and related resources development in the Mekong Basin.
(2) To learn about different approaches for water and related resources development and basin management in order to support the implementation of IWRM in the Mekong Basin.
(3) To learn about different legal/administrative enabling mechanisms for transboundary basin development and management to further enhance the implementation of the 1995 Mekong Agreement, with a view to maximizing the opportunities of the LMB countries to realize their individual and collective development aims.[701]

What cannot be concealed in the summary report of this "high level study visit" is that the Columbia River Basin "[i]s also one of the most powerful rivers in the world in terms of hydroelectric development ... with 14 dams on the mainstem, 219 major dams on tributaries, and more than 1,000 small dams scattered throughout the basin," and the prevailing interests of the LMRB states.[702] Perhaps the ultimate irony is the choice of this river basin for the study: one with a treaty which does not comprehensively deal with all of its tributaries. The author asserts that if the strategies observed in the MRC's Columbia River study visits are used by the MRC as a model for the LMRB, there can be no doubt about the agenda of further steps: tributaries in the LMRB will be transformed.

700 MRC, "High Level Study Visit to the Columbia River Basin: Back-to-Office Report" (MRC Basin Development Plan Programme, Phase 2, June 2008) 1.

701 *Id.* at 2–3 (emphasis added).

702 Osborn, *supra* note 698, at 1.

3.2.3 *A Community of Interest in Mekong Tributaries? The Influence of the* River Oder *and the* Gabcikovo-Nagymaros *(Hungary/Slovakia) Cases*

The 1929 *River Oder* case,[703] a landmark decision in tributary watercourse law, is dispositive of whether the jurisdiction of the International Commission of the River Oder and the freedom of navigation extended to the sections of that river's tributaries, the Warthe (Warta) or the Netze (Notec).[704] Prior to this case, a key question in international water law was whether tributaries of a river should be considered part of a river system. The Oder River in Western Europe was internationalized by Articles 331 and 332 of the 1919 Treaty of Versailles. Thereafter, the International Commission of the River Oder was established consisting of the riparian and nonriparian states which included UK, Czechoslovakia, Denmark, France, Germany, Sweden, and Poland. All members of the Commission asserted jurisdiction over navigable sections of the river including the Warta and Netze tributaries located wholly within Poland. Article 331 provides that all navigable sections of the river and its tributaries form a "system," including the source to the mouth. In the *River Oder* case, the issue was whether the jurisdiction of the Commission extended to the tributaries.

Predicated on the Court's interpretation of Article 331[705] derived from the "principles governing international fluvial law in general," the PCIJ stated that there was a "community of interest" in transboundary watercourses and held that:

> [W]hen consideration is given to the manner in which States have regarded the concrete situations arising out of the fact that a single waterway traverses or separates the territory of more than one State, and the possibility of fulfilling the requirements of justice and the considerations of utility which this fact places in relief, it is at once seen that a solution of the problem has been sought not in the idea of a right of passage in favour of upstream States, but in that of a community of interest of riparian States. This community of interest in a navigable river becomes the basis of a common legal right, the essential features of which are the perfect equality of all riparian States in the use of the whole course of the river and the exclusion of any preferential privilege of any one riparian State in relation to the others.[706]

703 *River Oder* case, *supra* note 488.

704 See note 489, *supra.*

705 See note 490, *supra.*

706 *River Oder* case, *supra* note 488, PCIJ Ser. A No. 23, at 26–28.

The concept of a community of interest in the water derives from "the natural, physical unity of a watercourse."[707] The PCIJ held that the jurisdiction of the International Commission of the River Oder extended to sections of the two tributaries situated in Polish territory. The PCIJ did not suggest that the community of interest concept is limited only to navigation.[708] Lipper, in his discourse on equitable utilization, noted that "both [the Court's] language and its reasoning make [the PCIJ's passage quoted above] equally applicable to non-navigational uses."[709] Further, Lammers argues that:

> For forms of use other than navigation, the legal notion of community of interest could not, of course, find exactly the same application. As appears from the practice of States, each riparian State may make such other use of water only within the limits of its own territory. [However,] the other elements mentioned such as "the perfect equality of all riparian States" and "the exclusion of any preferential privilege of any one riparian State in relation to other" would in the line of the Court's thinking, probably apply similarly to water uses other than navigation.[710]

This assessment was confirmed by the ICJ in the *Gabcikovo-Nagymaros* case, as seen below. The interests surrounding the concept of a community of interest are "identified and safeguarded on the basis of equity."[711]

The 1997 judgment in the *Gabcikovo-Nagymaros* case (Danube River) confirmed the creation of a body of legal principles governing transboundary water resources, demonstrated the ICJ's adherence to the emerging concept of a community of interest in international watercourses, and summarized the current state of development of international water resources law. Rising in the Black Forest, the Danube, the second longest river in Europe, flows eastward 2,860 km across the borders of nine countries before discharging into the Black Sea via the Danube delta in Romania and the Ukraine.[712] The basic facts of the *Gabcikovo-Nagymaros* case are these:[713] Czechoslovakia and Hungary

---

707 McCaffrey, *supra* note 9, at 148 (citations omitted).

708 *Id.* at 150.

709 Jerome Lipper, "Equitable Utilization" in A. Garretson et al. (eds.), *The Law of International Drainage Basins* (Oceana Publications, 1968) 29.

710 J. Lammers, *Pollution of International Watercourses* (Matinus Nijhoff Publishers, 1984) 507.

711 Owen McIntyre, *Environmental Protection of International Watercourses under International Law* (Ashgate Publishing Ltd., 2007) 30.

712 *Gabcikovo-Nagymaros* case, *supra* note 474, at 18, para 16.

713 For an in-depth analysis of the *Gabcikovo-Nagymaros* case, see McCaffrey. *supra* note 9, at 210–221.

entered into a treaty[714] on 16 September 1977 (in force on 30 June 1978), for diversion of the Danube River, which formed the countries' common boundary for 142 km,[715] and to build a series of barrages or dams. The plans were to divert the river into a 30-km by-pass canal at Gabcikovo in Czechoslovakia. A structure downstream at Nagymaros was to generate hydroelectricity and moderate the flow of waters released. If not moderated, this flow could adversely impact water management and navigation between Gabcikovo and Nagymaros.

In response to public protests about the project's effects on water quality, water quantity in certain locales, biota and the Danube ecosystem, Hungary suspended all work on the project in May 1989. Anticipating Hungary's withdrawal from the project in the 1980s, Czechoslovakia made provisional plans under the name "Variant C," a proposed dam at Cunovo, in Czech territory.[716] Hungary attempted to unilaterally denounce the 1977 treaty on 19 May 1992, invoking grounds for termination under the 1969 Vienna Convention and declaring a state of necessity or grave and imminent peril.

In 1993, Slovakia became an independent state and successor to Czechoslovakia in this case. The ICJ considered whether Hungary's suspension and subsequent abandonment of the project could be justified by the defense of necessity.[717] The Court responded negatively, holding that environmental concerns did not amount to a necessity when Hungary could not establish a grave and imminent peril.[718] On the other hand, Slovakia's Variant C was also deemed an intentionally wrongful act.[719] Accordingly, the ICJ found that both parties had breached their international obligations.

The law of treaties is dispositive of whether a convention is or is not in force, and whether it has or has not been properly suspended or denounced. The ICJ observed that the 1969 Vienna Convention[720] defines the parameters by which a treaty may be lawfully denounced or suspended.[721] The law of state

714 Treaty concerning the Construction and Operation of the Gabcikovo-Nagymaros System of Locks, 32 ILM 1247 (1993) [hereinafter 1977 Gabcikovo-Nagymaros Treaty].

715 *Gabcikovo-Nagymaros* case, *supra* note 474, para. 16.

716 *Id.*, par. 23.

717 *Id.*, para. 27 and para. 40.

718 *Id.*, para. 54–59.

719 *Id.*, para 78.

720 1969 Vienna Convention, *supra* note 512.

721 *Gabcikovo-Nagymaros* case, *supra* note 474, para. 47.

responsibility[722] determines whether such action was wrongful.[723] In the *Gabcikovo-Nagymaros* case, the Court held that the treaty remained in force and Hungary remained obligated to Czechoslovakia to fulfill the terms.[724]

Hungary claimed that both impossibility of performance and fundamental changes of circumstances justified its 19 May 1992 termination of the treaty with Czechoslovakia.[725] The Court observed that the 1969 Vienna Convention was not directly applicable, as both states ratified it only after their 1977 treaty. The Court found that because Articles 60 to 62,[726] which relate to termination

---

722 Draft Articles on the *Responsibility of States for Internationally Wrongful Acts*, Report of the ILC on the Work of its Fifty-third Session, UN GAOR, 56th Sess., Supp. No. 10, p. 43, UN Doc. A/56/10 (2001).

723 *Gabcikovo-Nagymaros* case, *supra* note 474, para. 48.

724 *Id.*, para. 115.

725 *Id.*, para. 92.

726 1969 Vienna Convention, *supra* note 512, arts. 60–62. Article 60 of the 1969 Vienna Convention "Termination or suspension of the operation of a treaty as a consequence of its breach," provides:

"1. A material breach of a bilateral treaty by one of the parties entitles the other to invoke the breach as a ground for terminating the treaty or suspending its operation in whole or in part.

2. A material breach of a multilateral treaty by one of the parties entitles:

(*a*) the other parties by unanimous agreement to suspend the operation of the treaty in whole or in part or to terminate it either:

(i) in the relations between themselves and the defaulting State; or

(ii) as between all the parties;

(*b*) a party specially affected by the breach to invoke it as a ground for suspending the operation of the treaty in whole or in part in the relations between itself and the defaulting State;

(*c*) any party other than the defaulting State to invoke the breach as a ground for suspending the operation of the treaty in whole or in part with respect to itself if the treaty is of such a character that a material breach of its provisions by one party radically changes the position of every party with respect to the further performance of its obligations under the treaty.

3. A material breach of a treaty, for the purposes of this article, consists in:

(*a*) a repudiation of the treaty not sanctioned by the present Convention; or

(*b*) the violation of a provision essential to the accomplishment of the object or purpose of the treaty.

4. The foregoing paragraphs are without prejudice to any provision in the treaty applicable in the event of a breach.

5. Paragraphs 1 to 3 do not apply to provisions relating to the protection of the human person contained in treaties of a humanitarian character, in particular to provisions prohibiting any form of reprisals against persons protected by such treaties."

and suspension, declare customary international law, they still apply to bind the states irrespective of when they signed the Convention.[727] The 1977 treaty between the two states does not contain a termination provision.[728] Hungary also asserted a right to terminate the 1977 treaty based on a state of necessity;[729] however, the ICJ determined that was an insufficient ground for termination of a treaty and can only be invoked to exonerate failure to implement a treaty.[730] Indeed, cessation of the necessity triggers revival of the duty to implement a treaty.[731] Parties must terminate a treaty by mutual agreement unless one of the 1969 Vienna Convention grounds is applicable.[732]

Article 61, "Supervening impossibility of performance," provides:
"1.A party may invoke the impossibility of performing a treaty as a ground for terminating or withdrawing from it if the impossibility results from the permanent disappearance or destruction of an object indispensable for the execution of the treaty. If the impossibility is temporary, it may be invoked only as a ground for suspending the operation of the treaty.
2. Impossibility of performance may not be invoked by a party as a ground for terminating, withdrawing from or suspending the operation of a treaty if the impossibility is the result of a breach by that party either of an obligation under the treaty or of any other international obligation owed to any other party to the treaty."
Article 62, "Fundamental change of circumstances," provides:
"1.A fundamental change of circumstances which has occurred with regard to those existing at the time of the conclusion of a treaty, and which was not foreseen by the parties, may not be invoked as a ground for terminating or withdrawing from the treaty unless:
(*a*) the existence of those circumstances constituted an essential basis of the consent of the parties to be bound by the treaty; and
(*b*) the effect of the change is radically to transform the extent of obligations still to be performed under the treaty.
2.A fundamental change of circumstances may not be invoked as a ground for terminating or withdrawing from a treaty:
(*a*) if the treaty establishes a boundary; or
(*b*) if the fundamental change is the result of a breach by the party invoking it either of an obligation under the treaty or of any other international obligation owed to any other party to the treaty.
3. If, under the foregoing paragraphs, a party may invoke a fundamental change of circumstances as a ground for terminating or withdrawing from a treaty it may also invoke the change as a ground for suspending the operation of the treaty."

727 *Gabcikovo-Nagymaros* case, *supra* note 474, para. 99.
728 *Id.*, para. 100.
729 *Id.*, para. 40.
730 *Id.*, para. 101.
731 *Id.*
732 *Id.*

Hungary argued that pursuant to Article 61 of the 1969 Vienna Convention, impossibility of performance was applicable to terminate the treaty.[733] However, Article 61 requires the "permanent disappearance or destruction of an object indispensable for the execution of the treaty" to justify termination of a treaty on this ground. In the *Gabcikovo-Nagymaros* case, the parties had rejected an amendment to allow impossibility of performance based on serious financial difficulties.

Further defending its actions on grounds of fundamental change of circumstances (*rebus sic stantibus*), Hungary cited profound political changes (breakup of the Soviet Union), diminishing economic value, increased environmental knowledge, and the evolution of new norms and prescriptions of international environmental law as fundamental changes warranting treaty termination.[734] The Court did not find these arguments compelling, stating that the prevalent political conditions were not closely linked to the object and purpose of the treaty.[735] While the evolution and progression of international environmental law may have been unforeseen, the 1977 treaty's articles 15, 19, and 20[736] were included in the agreement to accommodate change.[737] For this to work, the existence of the circumstances at the time of the treaty's conclusion must have constituted an essential basis of the consent of the parties to be bound by the treaty.[738]

733 *Id.*, para. 102.

734 *Id.*, para. 104.

735 *Id.*

736 1977 Gabcikovo-Nagymaros Treaty, *supra* note 714, arts. 15, 19 and 20. Article 15, "Protection of Water Quality," provides:
"1. The Contracting Parties shall ensure, by the means specified in the joint contractual plan, that the quality of the water in the Danube is not impaired as a result of the construction and operation of the System of Locks.
2. The monitoring of water quality in connection with the construction and operation of the System of Locks shall be carried out on the basis of the agreements on frontier waters in force between the Governments of the Contracting Parties."
Article 19, "Protection of Nature," provides:
"The Contracting Parties shall, through the means specified in the joint contractual plan, ensure compliance with the obligations for the protection of nature arising in connection with the construction and operation of the System of Locks."
Article 20, "Fishing Interests," provides:
"The Contracting Parties, within the framework of national investment, shall take appropriate measures for the protection of fishing interests in conformity with the Danube Fisheries Agreement, concluded at Bucharest on 29 January 1958."

737 *Gabcikovo-Nagymaros* case, *supra* note 474, para. 104.

738 *Id.*

Hungary argued that Czechoslovakia breached the agreement by planning, constructing and operating Variant C.[739] The ICJ stated that a breach must be material to give rise to counter measures by an injured state.[740] In *Gabcikovo-Nagymaros*, the Court determined that Variant C did not constitute grounds for termination under the law of treaties. Hungary terminated the treaty on 19 May 1992, before Czechoslovakia violated the treaty by diverting the Danube waters in October 1992.[741] Because Slovokia, successor to Czechoslovakia, had not yet violated the 1977 treaty, Hungary's termination was premature.[742] Hungary maintained that both parties repudiated the treaty through their conduct.[743] The ICJ disagreed and responded that, although both states failed to comply with their obligations, the reciprocal wrongful actions did not effectively end the treaty or justify its termination, citing Article 26 of the 1969 Vienna Convention,[744] "*Pacta sunt servanda*," requiring the good faith observance of treaties.[745] Czechoslovakia resisted efforts to terminate after Hungary sent notice, the former state declaring the latter's notice to be without legal effect.[746] Indeed, Hungary failed to give sufficient notice of termination by offering only six days' notice.[747] The ICJ rejected this notice as insufficient because Hungary had suffered no injury.[748] The ICJ's judgment declared that Hungary's May 19th termination notice was ineffective; that Slovakia, as successor to Czechoslovakia, became a party to the treaty; that the states must negotiate in good faith to achieve the objectives of the treaty; that Hungary must compensate Slovakia for the damage it suffered after Hungary suspended and abandoned the project and that Slovakia must compensate Hungary for putting into effect its provisional solution (Variant C).[749]

While the exact contours of the "community of interest" theory are indeterminate, McCaffrey, who represented Slovakia in the *Gabcikovo-Nagymaros* case, observed that based on the ICJ's opinion "[i]t would, ... not be unreasonable to conclude that [the] considerations [discussed by the PCIJ in the

739 *Id.*, para. 105.

740 *Id.*, para. 106.

741 *Id.*, para. 108.

742 *Id.*

743 *Id.*, para. 114.

744 1969 Vienna Convention, *supra* note 512, art 26. Article 26 of the 1969 Vienna Convention, "*Pacta sunt servanda*," provides: "Every treaty in force is binding upon the parties to it and must be performed by them in good faith."

745 *Gabcikovo-Nagymaros* case, *supra* note 474, para. 114.

746 *Id.*

747 *Id.*, para. 109.

748 *Id.*

749 *Id.*, para. 155.

River Oder case], and thus the notion of community of interest, apply to uses of international watercourses other than navigation."[750] He observes that in the *River Oder* case "[t]he Court's reference to a *single waterway* implies that it had in mind the natural phenomenon of the unity of a watercourse that may run through or along different states due to the placement of political boundaries."[751] Indeed, McCaffrey's conclusion is supported by selected passages in the *Gabcikovo-Nagymaros* case in which the ICJ stated that:

> In 1929, the Permanent Court of International Justice, with regard to navigation on the River Oder, stated as follows:
>
> "[the] community of interest in a navigable river becomes the basis of a common legal right, the essential features of which are the perfect equality of all riparian States in the use of the whole course of the river and the exclusion of any preferential privilege of any one riparian State in relation to the others" (*Territorial Jurisdiction of the International Commission of the River Oder, Judgment No. 16, 1929, P.C.I.J., Series A, No. 23,* p. 27).
>
> Modern development of international law has strengthened this principle for non-navigational uses of international watercourses as well, as evidenced by the adoption of the Convention of 21 May 1997 on the Law of the Non-Navigational Uses of International Watercourses by the United Nations General Assembly.[752]

Thus, the community of interest doctrine was applied for the first time to *non-navigational* uses of international watercourses in the *Gabcikovo-Nagymaros* case. McCaffrey commented that the ICJ "breath[ed] new life into the powerful idea of 'community of interest', originally enunciated in the *River Oder* case."[753] Indeed, in the *Gabcikovo-Nagymaros* case, the Court used the community of interest theory to support its finding "that one riparian may not unilaterally assum[e] control of a shared resource."[754]

Analysis of these two important opinions reveals that the issues in these cases contributing to the development of tributary watercourse law are essentially threefold. First, the PCIJ decision in the *River Oder* case extends the freedom of navigation of an international watercourse to its tributaries. Prior to the *River Oder* case, some ambiguities existed as to whether tributaries were legally part of a river system. Second, neither case subordinates non-navigational uses of

750 McCaffrey, *supra* note 9, at 150 (citations omitted).

751 *Id.* at 149–150. (internal quotations omitted) (emphasis added).

752 *Gabcikovo-Nagymaros* case, *supra* note 474, para. 85.

753 McCaffrey, *supra* note 9, at 217.

754 *Id.*

international watercourses to any use preference below navigation. Third, and perhaps most importantly, both cases are dispositive of the existence of a community of interest in international watercourses and their tributaries. These opinions, in effect, constitute a judicial repudiation of absolute territorial sovereignty, recognize that international watercourses are shared resources, and have far-reaching implications for tributaries in the Mekong legal regime.

Few commentators presuppose that a principle of a community of interest in international watercourses "is likely soon to evolve" into a customary international law.[755] As McCaffrey notes, "[i]ts precise legal implications for non-navigational uses are less than completely clear."[756] Unquestionably, the concept of a "community of Mekong nations" in the 1995 Mekong Agreement and the "community of interest" theory aids in unveiling the circular and fallacious reasoning shrouded in Article 5(A). The community of interest doctrine is a powerful concept when applied to the LMRB states because planned measures by one state on a transboundary tributary could affect another state's use of a watercourse just as significantly as if that measure were applied to the mainstream.

3.2.4 Arizona v. California

State sovereignty over tributary watercourses has not escaped the scrutiny of the US Supreme Court, the very Court that fashioned the equitable apportionment doctrine, as evidenced by its interpretation of the 1928 Boulder Canyon Project Act[757] in *Arizona v. California*.[758] The central issue in that case concerned water allocation from the Colorado River and its tributaries.[759] Water use in this river basin is governed by perhaps the most convoluted and complex array of water rights of any basin in the world, contained in a web of applicable laws, compacts, court decrees, and treaties that govern each state's share of the water, known collectively as "the Law of the River."[760]

Rising from snow pack in the Rocky Mountains of the western United States, the Colorado cascades southward, flowing through Colorado, Wyoming, and Utah before entering into the Grand Canyon. There, the Upper Basin is separated from the Lower Basin arbitrarily at Lee Ferry by the compromise that

755 McIntyre, *supra* note 711, at 33.
756 McCaffrey, *supra* note 9, at 161.
757 1928 Boulder Canyon Project Act, *supra* note 13.
758 *Arizona v. California*, *supra* note 13.
759 *Id.* at 551.
760 McCaffrey, *supra* note 9, at 333.

is the 1922 Colorado River Compact.[761] Turning west, the Colorado flows turbulently through the deep gorges of the Grand Canyon, veers south in southern Nevada to form the border of the tristate region of Nevada, Arizona, and California, and empties what is left of its diminished flow (if any) into the Gulf of California after forming part of the US-Mexican border. As it plunges from the Rockies and trickles into the Gulf of California, along the way it receives tributary waters from Wyoming, Colorado, Utah, Nevada, New Mexico, and Arizona. The 2,334-km long Colorado and its tributaries drain a 640,000-km^2 basin—practically one-twelfth of the continental US excluding Alaska—which includes the most arid regions in that country.[762]

The 1922 Colorado River Compact negotiated by the seven states and the US divides the entire basin into two parts, the Upper Basin and the Lower Basin, separated geographically at a point on the river in Arizona known as Lee Ferry. Article III(a)[763] of the 1922 Compact allocates to each basin in perpetuity 7.5 million acre-feet of water annually from the "Colorado River System," defined in Article II(a) as "the Colorado River *and its tributaries* within the United States of America."[764] (emphasis added). Largely because of its failure to allocate each state's share of the water resources, the 1922 Compact failed to live up to the expectations of Congress. Although it resolved issues between the two basins, it failed to quiet water allocation rivalries between the Lower Basin states. Consequently, Arizona, fearing that the law of prior appropriation[765] in California

761 Signed at Santa Fe, New Mexico, 24 November 1922, H. Doc. 605, 67th Cong., 4th Sess., pp. 8–12 (1923); 70 Cong. Rec. 324 (1968). [hereinafter 1922 Colorado River Compact or 1922 Compact]. The 1922 Colorado River Compact can also be found in US Dept. of the Interior, Documents on the Use and Control of the Waters of Interstate and International Streams 39 (1956).

762 *Arizona v. California, supra* note 13, at 552; National Research Council of the National Academies, *Colorado River Basin Water Management: Evaluating and Adjusting to Hydroclimatic Variability* (National Academies Press, 2007) 13.

763 1922 Colorado River Compact, *supra* note 761, art. III(a). Article III(a) of the 1922 Colorado River Compact provides: "There is hereby apportioned from the Colorado River System in perpetuity to the Upper Basin and to the Lower Basin, respectively, the exclusive beneficial consumptive use of 7,500,000 acre-feet of water per annum, which shall include all water necessary for the supply of any rights which may now exist."

764 1922 Colorado River Compact, *supra* note 761, art. II(a). Article II(a) of the 1922 Colorado River Compact states: "The term 'Colorado River System' means that portion of the Colorado River and its tributaries within the United States of America."

765 *See Irwin v. Phillips,* 5 Cal. 140 (1855). Prior appropriation provides that an individual who first appropriates and uses water for a reasonable beneficial purpose has a right which is superior to that of later appropriators. The prior appropriation doctrine of "first in time first in right" in California injected uncertainty into 1922 Compact negotiations because

would provide the latter state a formula for allocation of a lion's share of the water resources in the Lower Basin to fuel development, vehemently protested inclusion of tributaries, especially the Gila River, in the 1922 Compact's allocation scheme and refused to ratify it. In an attempt to secure ratification of the 1922 Compact, the governors of all seven basin states met in Denver in 1925 and again in 1927. A proposal was put forth to apportion the Lower Basin's 7.5 million acre-feet of water required in the 1922 Compact as follows: 300,000 acre-feet to Nevada, 3 million acre-feet to Arizona, 4.2 million acre-feet to California, and all unapportioned water would be subject to equal reapportionment between Arizona and California after 1963. Each state would have the exclusive beneficial consumptive use of such tributaries within its boundaries before those tributaries empty into the mainstream, except that Arizona tributary waters in excess of 1 million acre-feet could, under some circumstances, be subject to diminution by reason of a US treaty with Mexico. This proposal foundered as representatives from California countered with a demand for 4.6 million acre-feet and continued to do their compassionate best to quench the thirst felt so strongly in that state, and Arizona, clinging fervently to state sovereignty, insisted on exemption of tributaries, particularly the Gila, from any apportionment.[766] Finally, on 21 December 1928, Congress passed the Boulder Canyon Project Act, a piece of legislation that appropriated funds for a megaproject, the Hoover dam, at Boulder Canyon, on the Arizona-Nevada border.

The text of the 1928 Boulder Canyon Project Act, *inter alia*, restricted allocation to California to "water of and from the Colorado River," not from the Colorado River System.[767] This legislation provided for negotiation of an agreement

---

California had appropriated and used Colorado River water before the upper States. The fear caused by this uncertainty was not limited to the northernmost states. Arizona was equally fearful that prior appropriation would undermine a just allocation of water in the Lower Basin.

766 Hearings on H.R. 5773 before the House Committee on Irrigation and Reclamation (United States) 70th Cong., 1st Sess., 1927, at 30–31 and 402–205.

767 1928 Boulder Canyon Project Act, *supra* note 13, sec. 4(a). Section 4(a) provides: "This Act shall not take effect and no authority shall be exercised hereunder and no work shall be begun and no moneys expended on or in connection with the works or structures provided for in this Act, and no water rights shall be claimed or initiated hereunder, and no steps shall be taken by the United States or by others to initiate or perfect any claims to the use of water pertinent to such works or structures unless and until (1) the States of Arizona, California, Colorado, Nevada, New Mexico, Utah, and Wyoming shall have ratified the Colorado River compact, mentioned in Section 13 hereof, and the President by public proclamation shall have so declared, or (2) if said States fail to ratify the said compact within six months from the date of the passage of this Act then, until six of said States, including the State of California, shall ratify said compact and shall consent to waive the provisions of the first paragraph of Article XI of said compact, which makes the same binding and

which limited California to 4.4 million acre-feet of water, allocated the remaining water resources to Arizona (2.8 million acre-feet) and Nevada (300,000 acre-feet), established the Federal Government (Secretary of the Interior through the Bureau of Reclamation) as the river basin manager, and authorized the construction, operation, and maintenance of a megaproject (dam and attendant works) in the mainstream.[768] The agreement was also to allow Arizona exclusive use of the Gila River, free and clear of Mexican interests. The

---

obligatory only when approved by each of the seven States signatory thereto, and shall have approved said compact without conditions, save that of such six-State approval, and the President by public proclamation shall have so declared, and, further, until the State of California, by act of its legislature, shall agree irrevocably and unconditionally with the of the States of Arizona, Colorado, Nevada, New Mexico, Utah, and Wyoming, as an express covenant and in consideration and Wyoming, as an express covenant and in consideration of the passage of this Act, that the aggregate annual consumptive use (diversions less returns to the river) *of water of and from the Colorado River* for use in the State of California, including all uses under contracts made under the provisions of this Act and all water necessary for the supply of any rights which may now exist, shall not exceed four million four hundred thousand acre-feet of the waters apportioned to the lower basin States by paragraph (a) of Article III of the Colorado River compact, plus not more than one-half of any excess or surplus waters unapportioned by said compact, such uses always to be subject to the terms of said compact. The States of Arizona, California, and Nevada are authorized to enter into an agreement which shall provide (1) that of the 7,500,000 acre-feet annually apportioned to the lower basin by paragraph (a) of Article III of the Colorado River compact, there shall be apportioned to the State of Nevada 300,000 acre-feet and to the State of Arizona 2,800,000 acre-feet for exclusive beneficial consumptive use in perpetuity, and (2) that the State of Arizona may annually use one-half of the excess or surplus waters unapportioned by the Colorado River compact, and (3) that the State of Arizona shall have the exclusive beneficial consumptive use of the Gila River and its tributaries within the boundaries of said State, and (4) that the waters of the Gila River and its tributaries, except return flow after the same enters the Colorado River, shall never be subject to any diminution whatever by any allowance of water which may be made by treaty or otherwise to the United States of Mexico but if, as provided in paragraph (c) of Article III of the Colorado River compact, it shall become necessary to supply water to the United States of Mexico from waters over and above the quantities which are surplus as defined by said compact, then the State of California shall and will mutually agree with the State of Arizona to supply, out of the main stream of the Colorado River, one-half of any deficiency which must be supplied to Mexico by the lower basin, and (5) that the State of California shall and will further mutually agree with the States of Arizona and Nevada that none of said three States shall withhold water and none shall require the delivery of water, which cannot reasonably be applied to domestic and agricultural uses, and (6) that all of the provisions of said tri-State agreement shall be subject in all particulars to the provisions of the Colorado River compact, and (7) said agreement to take effect upon the ratification of the Colorado River compact by Arizona, California, and Nevada." (emphasis added).

768 *Arizona v. California*, *supra* note 13, at 560–561.

three states never entered into any agreements. Arizona ratified the 1922 Colorado River Compact in 1944,[769] and later, in 1963, sued California challenging the water allocations contained in the 1928 Boulder Canyon Project Act.

In the instant case, Arizona claimed that the apportionment of Colorado River waters in the 1928 Boulder Canyon Project Act included only mainstream waters. California, seeking more water to support growth and development, made a water grab for Arizona's Gila River and attempted to use an interplay of terms between the 1928 Boulder Canyon Project Act ("water of and from the Colorado River") and the 1922 Compact, contending that the apportionment included all tributary waters of the "Colorado River System" as provided in the latter document.[770] The Special Master appointed by the Court concluded that the 1922 Compact, the law of prior appropriation, and equitable apportionment did not control the issues.[771] The Court agreed because in this case Congress had made a comprehensive statutory apportionment in the 1928 Boulder Canyon Project Act.[772] In the Court's opinion, Justice Black stated that "[C]ongress, in passing the [Boulder Canyon] Project Act, intended to and did create its own comprehensive scheme for apportionment among California, Arizona and Nevada of the Lower Basin's share of the mainstream waters of the Colorado River, leaving each state its tributaries."[773] The Court interpreted Section 4(a) of the 1928 Boulder Canyon Project Act to limit California to "water of and from the Colorado River, not of and from the 'Colorado River System'", and determined that legislative history was dispositive of the issue of reservation of tributaries to each state.[774] This determination was a huge blow to California and its view that the apportionment by Congress included both tributary and mainstream waters and not just water from the mainstream.[775] The Supreme Court held that Congress, in the 1928 Boulder Canyon Project Act, intended to allocate only mainstream waters and that tributaries belonged to the individual states, shattering the drainage basin concept and thwarting any hopes for holistic management of the Lower Basin.

Exclusion of tributaries from the 1928 Boulder Canyon Project Act in the Lower Basin is in stark contrast to the 1948 Upper Colorado River Basin Compact,[776] which provides an interesting approach to tributary watercourse law, and includes subsidiary agreements called sub-compacts that apportion

769 ARIZ. REV. STAT. Secs. 645–1311 (2006).

770 *Arizona v. California*, *supra* note 13, at 563.

771 *Id.* at 562.

772 *Id.* at 565.

773 *Id.* at 564–565.

774 *Id.* at 568.

775 *Id.* at 567.

776 63 Stat. 31 (1949).

water use in tributaries.[777] Vick cogently summarizes the substantive and procedural framework of the sub-compacts and the La Plata River Compact, which together cover water allocation on Upper Basin tributaries:

> Article XI of the 1948 Upper Basin Compact is the sub-compact for the Little Snake River and its tributaries apportioning its waters between the states of Colorado and Wyoming. The states agree to administer this tributary using a system of interstate priorities without regard to the state of use, except in times of curtailment which results from a shortage of supply within the basin. In the event of shortage such that upper basin curtailment is required it is agreed that, to the extent possible, water use from the Little Snake River will be curtailed on an equal basis between the states. The parties agree that the Little Snake River is administered by each state according to its water laws.
>
> Article XII of the Upper Basin Compact is a sub-compact between the states of Utah and Wyoming apportioning the use of waters from the Green River and its tributaries. Existing uses at the time of the Compact are apportioned by priority date without regard to state boundaries. Uses initiated after the effective date of the Compact are apportioned 50% to each state. This sub-compact is administered by a Special Water Commissioner who is jointly appointed by the State Engineers of each state.
>
> Article XIII of the Upper Basin Compact is a sub-compact between the states of Colorado and Utah for apportionment of the Yampa River, a tributary to the Green River. The upstream state of Colorado agrees not to deplete the flow of the Yampa River below "an aggregate of 5,000,000 acre-feet for any period of ten consecutive years" as measured at a designated guaging station. No additional administrative institution is created.
>
> The San Juan River is apportioned between the states of Colorado and New Mexico in Article XIV of the Upper Basin Compact. The state of New Mexico is not riparian to the mainstream of the Colorado River but does supply water to the San Juan River tributary which runs through the northwest portion of the state. This sub-compact provides that the upstream state of Colorado "agrees to deliver to the State of New Mexico from the San Juan River and its tributaries which rise in the State of Colorado a quantity of water which shall be sufficient, together with water originating in the San Juan Basin in the State of New Mexico, to enable the State of New Mexico to make full use of the water apportioned to the State of New Mexico by Article III of the [Upper Basin] Compact", its 11.25% share. All uses as of the effective date of the Compact

777 Vick, *supra* note 450, at 150–154.

> are recognized with a prior right as are "uses of water contemplated by projects authorized." The technical determinations required to make this allocation are made by the Upper Basin Commission. It is reported that New Mexico would not approve the Upper Basin Compact without protection for use of its full 11.25% share from the San Juan River.
>
> Article X confirms and incorporates the 1922 La Plata River Compact between the states of Colorado and New Mexico apportioning this tributary. In the 1922 La Plata River Compact, the State of Colorado agrees to establish and maintain two guaging stations on the La Plata River, one upstream near the community of Hesparus and one downstream near the boundary with New Mexico. The Compact allocates to New Mexico one half the mean flow as measured at the Hesparus guaging station. However, when the "flow of the river is so low that … the greatest beneficial use of its waters may be secured by distributing all of its waters successively to the lands in each state in alternating periods … the use of the waters may be so rotated between the two states …". This scheme attempts to preserve uses within the downstream state when half the flow is not sufficient to convey the water and provide for its beneficial use. This Compact is administered by the Colorado and New Mexico State Engineers.[778]

These tributary sub-compacts and the La Plata River Compact, nested within the 1948 Compact, provide a holistic approach to Upper Basin water resources management and governance if "granted only," as Vick points out, in "the upper basin."[779] Noteworthy is that these bilateral sub-compacts ensure that tributary water allocations meet local stakeholder needs and are negotiated "closer to the user level and only by those states affected."[780] Vick posits that "For natural systems such as watercourses, the tributary sub-basin may be the lowest effective level of governance."[781] The LMRB states would do well to mark this state of play between the master compact and nested subsidiary agreements and consider a similar concept in the Mekong legal regime.

3.2.5 *The Nigeria-Niger Tributary Agreement*

Like the Little Snake, Green, Yampa, and San Juan rivers addressed in the aforementioned sub-compacts, and the La Plata River in its Compact in the

778 *Id.* at 150–152 (internal citations omitted).

779 *Id.* at 150.

780 *Id.* at 152.

781 *Id.* at 153.

Upper Colorado River Basin,[782] some Niger Basin tributaries are addressed by a subsidiary agreement.[783] The 1990 Nigeria-Niger Tributary Agreement governs "[t]he equitable development, conservation and use of the water resources which are bisected by, or form, the common frontier between the Contracting Parties [of the basin states of Nigeria and Niger]."[784] This tributary agreement provides that the parties are mindful of their commitments under the larger Niger Basin agreement framework, and is subsidiary to the 1980 Convention Creating the Niger Basin Authority.[785]

The Niger, the principal river of western Africa and third longest on that continent, rises in the Guinea Highlands, flows through Mali and Niger, forming the border between the latter state and Benin, then courses through Nigeria and ultimately discharges into the Atlantic Ocean via the massive Niger delta. The Niger is 4,200 km long and drains a basin of 2,270,000 km^2 in area.[786] This basin encompasses parts of nine of the poorest countries in the world and is home to 100 million people.[787] As Wormser aptly notes, the basin "is a vital, complex asset ... [and] is not simply water, but is ... [like the Mekong] an origin of identity, a route for migration and commerce, a source of potential conflict, and a catalyst for cooperation."[788]

Although the 1990 Nigeria-Niger Tributary Agreement memorializes the need for the contracting parties to remain "mindful"[789] of their commitments under the broader Niger Basin treaty, this tributary agreement, like the sub-compacts in the Upper Colorado River Basin, underscores the need for subsidiary agreements at a later stage in the equitable development, conservation, and use of

782 *Id.* at 150–152.

783 The shared river basins of the Maggia/Lamido, Gada/Goulbi of the Maradi basin, Tagwai/El Fadama and the lower section of the Komadougou-Yobe are covered in the Agreement Between the Federal Republic of Nigeria and the Republic of Niger Concerning the Equitable Sharing in the Development, Conservation and Use of Their Common Water Resources, 18 July 1990 [hereinafter 1990 Nigeria-Niger Tributary Agreement], available at the Transboundary Freshwater Dispute Database [hereinafter TFDD] at <http://ocid.nacse.org/tfdd/tfdddocs/483ENG.pdf> (accessed June 16, 2017).

784 1990 Nigeria-Niger Tributary Agreement, *supra* note 783, art.1.

785 Convention Creating the Niger Basin Authority signed at Faranah on 21 November 1980, available at TFDD at <http://ocid.nacse.org/tfdd/treaties.php> (accessed 16 June 2017).

786 Inger Andersen et al., *The Niger River Basin* (Katherin George Golitzen, ed., The World Bank, 2005) 11; Gupta, *supra* note 31, at 3.

787 Andersen et al., *supra* note 786, at x, 1 and 4–7. The Niger Basin traverses Benin, Burkina, Faso, Cameroon, Chad, Cote d'Ivoire, Guinea, Mali, Niger and Nigeria.

788 Michel Wormser, "Foreword" *in* Andersen et al., *The Niger River Basin: A Vision for Sustainable Management* (The World Bank, 2005) vii.

789 1990 Nigeria-Niger Tributary Agreement, *supra* note 783, prbl.

arterial watercourses. Indeed, subsidiary agreements would seem particularly relevant in legal regimes such as the Mekong, in which tributaries are marginalized or the extent of the basin is undefined. Andersen et al. discuss the need for subsidiary agreements in the Niger Basin legal regime and observes that:

> In view of the hydrological complexity of the Niger Basin, it is possible that subsidiary agreements among a subset of Basin countries will be required for specific river basin developments, which might not involve, or affect, all the basin countries. For example, developments on the Niger River tributaries in Chad or Cameroon will have no impact on Guinea, Mali or Niger and a subsidiary agreement between or among the involved and affected parties would therefore make sense. Such subsidiary agreements should of course fit within the framework of legal principles and agreements that exist at the level of the full river basin and to which the nine Basin countries have agreed.[790]

Subsequent agreements such as the 1990 Nigeria-Niger Tributary Agreement and the Upper Colorado River Basin sub-compacts, then, may be viewed as examples of the principle of subsidiarity, described by Kiss and Shelton as "a general organizing principle of governance, express[ing] a libertarian value in favor of making decisions and implementing them at the lowest effective level of government or other organization."[791] In her monumental work on integrated water law, Vick comments that "[l]ocal conditions within … tributary sub-basins impact, if not determine" water uses and allocation schemes "at the interstate and international level."[792] Perhaps like politics and counterinsurgency[793] then, all tributary water allocation is local.

790 Andersen et al., *supra* note 786, at 62.

791 Kiss and Shelton, *supra* note 214, at 800–801 (2004).

792 Vick, *supra* note 450, at 153. Prior to planned measures on transboundary tributaries, all relevant factors and circumstances should be evaluated including, but not limited to: (1) geographic, hydrographic, hydrological, climatic, ecological and other factors of a natural character; (2) the social and economic needs of the watercourse States concerned; (3) the population dependent on the watercourse in each watercourse State; (4) the effects of the use or uses of the watercourses in one watercourse State on other watercourse States; (5) existing and potential uses of the watercourse; (6) conservation, protection, development and economy of use of the water resources of the watercourse and the costs of measures taken to that effect; and (7) the availability of alternatives, of comparable value, to a particular planned or existing use. *See* 1997 UN Convention, *supra* note 18, art. 6.

793 Thomas H. Johnson and M. Chris Mason, *All Counterinsurgency Is Local* (October 2008) The Atlantic 36.

In Articles 3–5, the 1990 Nigeria-Niger Tributary Agreement establishes the Joint Commission for Cooperation, the body charged with determining the equitable shares of the water resources.[794] Existing uses of water on the tributaries are grandfathered in and take precedence over future uses.[795]

Preferences for competing uses are covered by Article 7, which declares:

---

794 1990 Nigeria-Niger Tributary Agreement, *supra* note 783, arts. 3 and 5. Article 3 provides: "1. The Contracting Parties undertake to collect, process and provide at regular intervals the Nigeria-Niger Joint Commission for Co-operation with all the data and information which, in the opinion of the Commission, are needed to arrive at equitable sharing determinations, and to monitor the continued viability thereof.
2. At the request of the Commission, the Contracting Parties shall:
(a) install in their territory the required measuring equipment, and protect such equipment from interference; and
(b) permit and facilitate inspections by the Commission of such equipment.
Article 4 provides: The Contracting Parties undertake to inform and consult with the Nigeria-Niger Joint Commission for Cooperation in advance of undertaking any project, programme or plan for the implementation of agreed upon equitable sharing determinations, or likely to have an appreciable impact on any such determination."
Article 5 declares:
"1. In determining the equitable share to which each Contracting Party is entitled pursuant to Article 2, the following factors shall be taken into account:
(a) the climate of the region, and its influence on rainfall patterns;
(b) rainfall patterns, and their influence on surface hydrology, and related hydrogeology;
(c) surface hydrology and related hydro-geology;
(d) existing uses of the waters;
(e) reasonable planned water development requirements;
(f) the economic and social development needs of the Contracting Parties;
(g) the dependence of local populations on the waters in question for their own livelihood and welfare;
(h) the availability of alternative sources of water to satisfy competing water demands;
(i) the practicability of compensations either in cash or in kind[,] one or the other[, to the] Contracting Party as a means of adjusting competing water demands;
(j) maintaining an acceptable environmental balance in and around a particular body of water;
(k) the avoidance of unnecessary waste in the utilization of waters, with due regard for the technological and financial capabilities of each Contracting Party;
(l) the proportion in which each Contracting Party contributes to the water balance of the basin.
2. Each factor is to be given the weight warranted by the circumstances peculiar to each individual river basin, or group of basins, and all factors so weighted are to be considered together and a determination arrived at on the basis of the whole."

795 1990 Nigeria-Niger Tributary Agreement, *supra* note 783, art.6. Article 6 provides: "A water use existing at the time an equitable sharing determination is made shall take precedence

> A water use existing at the time an equitable sharing determination is made shall take precedence over a competing existing use which came into being later in time, provided that:
> (a) the use prior in time is beneficial to both Contracting Parties and reasonable under the circumstances; and
> (b) the weighting of the factors under Article 5 does not warrant accommodation of the later use, in whole or in part.

Like the Upper Colorado River Basin sub-compacts, the 1990 Nigeria-Niger Tributary Agreement is an interesting one because it provides a subsidiary agreement to manage tributaries, furthers the goals of basin-wide governance, and holistic water resources management, and restricts management of tributaries to the affected basin states.

3.3 Chapter 3 Conclusion

In the 1990s, the powerful response to renewed nationalism in the LMRB by the Thai and Vietnamese resulted in the protracted negotiations that led to the effectuation of Article 5, the controversial provision on water usage in the 1995 Mekong Agreement. This water usage rule emanated from a compromise to accommodate the specific national interests of Thailand, rather than from a unified vision of sustainable development to protect the Mekong ecosystem. In this sense, Article 5(A), the "tributary rule," reflects the transition from interdependency of the former Mekong legal regime to the unbridled pursuit of national agendas under the umbrella of the 1995 Mekong Agreement. Article 5(A) is transformative and is turning LMRB tributaries into "contested waterscapes."[796] Water rights and interests are the bases for the distinction between the mainstream and tributaries, which form a hydrological perspective that is artificial, even nonsensical. Article 5(A) also presupposes that tributaries are intrastate rather than transboundary, a fiction easily refuted by the 3S Rivers—the Sekong in the Lao PDR and Cambodia, and the Sesan and Srepok in Vietnam and Cambodia.

---

over a future use, provided the existing use is beneficial to both Contracting Parties and reasonable under the circumstances."

796 *See generally*, Francois Molle et al., "Contested Mekong Waterscapes: Where to Next?" in Francois Molle et al. (eds.), *Contested Waterscapes in the Mekong Region: Hydropower, Livelihoods and Governance* (Earthscan, 2009) 383–413. (describing the impacts of contested waterscapes in the Mekong region).

Analysis of selected international and interstate legal regimes and case law reveals the disparate treatment of tributaries when states cling fervently to national sovereignty. The 1909 Boundary Waters Treaty evidences a qualified exclusion of tributary waters, albeit one tempered by the remedies for transboundary harm in Article II. Nevertheless, the 1909 Boundary Waters Treaty does not include basins and tributary sub-basins under one umbrella agreement although it can be observed that there is some degree of interdependence preserved by limiting unilateral acts to the extent that they cause transboundary harm. The exclusion of tributaries in the 1909 Boundary Waters legal regime did not work and these river arteries have been finessed into state practice.

The 1961 Columbia River Treaty does not deal with all the tributaries of that river basin. The Treaty does not mention the Okanagan, a major Canadian tributary which occupies a sub-basin with a rapidly increasing population. This is an unqualified exclusion of an important arterial river in a basin with the most dammed tributaries in the world. By lopping off and pruning many of its major tributaries, this treaty is, in most respects, the antithesis of what forward-thinking basin-wide river governance should entail, although it is lauded as a model for hydropower development.

The US Supreme Court's interpretation of the 1928 Boulder Canyon Project Act is a fascinating exposé on the state-centric water allocation in the tributaries of Lower Colorado River Basin. In *Arizona v. California*, the Court concluded that Congress, when it enacted the Boulder Canyon Project Act in 1928, "intended to and did create its own comprehensive scheme for the apportionment among California, Arizona, and Nevada of the Lower Basin's share of the mainstream waters of the Colorado River, leaving each state its tributaries."[797] Thus, the legislative and litigation options of water allocation have met with much the same results as international agreements in their disparate treatment of tributaries: Basin-wide governance of the Colorado River has been supplanted by state sovereignty. The Lower Colorado River Basin remains dissevered.

In contrast to the 1928 Boulder Canyon Project Act in the Lower Basin, tributaries in the Upper Basin are managed by sub-compacts nested within the 1948 Compact. This "subsidiary" approach to tributary water allocation and diversions is also evident in the 1990 Nigeria-Niger Tributary Agreement. Subsidiary agreements provide an interesting approach to tributary water allocations and diversions with a tripartite advantage: They further the goals of basin-wide governance and holistic water resources management, foster decision-making

797 *Arizona v. California*, *supra* note 13, at 564–565.

at the lowest level of government involving local stakeholders, and restrict management of tributaries to the affected basin states.

The *River Oder* and *Gabcikovo-Nagymaros* cases confirm a "community of interest" in international watercourses, and thus in a river's tributaries, especially those of a transboundary nature. While any predictions that this theory may evolve into a customary norm are probably exaggerations, the concept of a community of interest evidences a strong case for inclusion of tributaries in the Mekong hydropolitical equation. Indeed, this theory really informs other customary norms, mainly equitable and reasonable utilization. Sub-agreements for water usage and diversions on tributaries nested under the umbrella of the 1995 Mekong Agreement could provide a functional approach for concomitant sub-basin and basin management (albeit if only in the LMRB) and foster a "community of interest" in the Mekong River system. However, given the geopolitics and the protracted negotiations involved in drafting and reaching agreement on the 1995 Mekong Agreement, the same objective of adopting a basin-wide conjunctive management approach and demarginalizing tributaries in the LMRB could be accomplished by Cambodia, Lao PDR, and Thailand joining the UN Watercourses Convention, now that the latter treaty has entered into force and Vietnam has ratified it.

CHAPTER 4

Mekong Tributaries: Selected Case Studies

In the past two decades, water diversions on tributaries have moved up on the national agenda of most Mekong countries. Across the LMRB, flow alterations and diversions on tributaries have led to the depletion of stream flow-reliant ecosystems. The Lao PDR, Thailand, and Vietnam have extensively developed dams and irrigation infrastructure. The following two case studies underscore the disparate treatment of tributaries compared with the mainstream in the Mekong legal regime and highlight the ways in which national interests shape transboundary water governance in the LMRB. Drawing on the current dispute between Vietnam and Cambodia over hydropower development on the Sesan River at Yali Falls dam[798] and the controversy over the "Thai water grid,"[799] both of which severely impact Mekong tributaries, it is apparent that development concerns trump sustainability in the LMRB. These two projects epitomize the concept of unsustainable basin development,[800] and demonstrate the gap in what is politically announced in the 1995 Mekong Agreement and the realpolitik in the LMRB. As such, both cases remain controversial. Unfortunately, neither case has prompted real action by the MRC to engage in constructive resolution of the emerging international environmental challenges or to significantly enhance relations with stakeholders and community groups. Noteworthy is that both projects have been justified by national interests and state sovereignty.

4.1 Hydropower Development on the Sesan River

The Mekong tributaries' days of flowing unchecked look numbered and for some the days are long gone. As renowned water journalist Fred Pearce aptly notes, "The Mekong and *its tributaries* have been a mouthwatering prospect

798 *See generally,* Dangerous Waters, *supra* note 16; Down River, *supra* note 16.

799 *See generally,* Molle and Floch (2007) and Molle and Floch (2008a), *supra* note 14.

800 Adam S. Rix, "The Mekong River Basin: A Resource at the Cross-Roads of Sustainable Development" (2003) 21 *Temp. Envtl. L. and Tech. J.* 103, 104 (citing Susan Y. Noe, "Belize's Chalillo Dam: A Sustainability Analysis of a Proposed Development Project" (unpublished manuscript, April 4, 2002) 25.

 | DOI 10.1163/9789004362604_006

for hydroelectric engineers for half a century."[801] Unbridled hydropower development on Mekong transboundary tributaries has resulted in severe cross-border effects on indigenous peoples and basin ecology and has tested the efficacy of the Mekong legal regime to promote long-term environmental, social, and economic benefits.[802] Emblematic of this development is the construction and operation of the Yali Falls Dam, still causing friction between Vietnam and Cambodia on the Sesan River, an important Mekong tributary flowing through both states. Predictably, the dam caused severe impacts on the river and local villages, in some cases tens of kilometers downstream. The dam created a flashpoint of conflict between the two countries, which pointedly highlights the disparate treatment between the mainstream and tributaries in the Mekong legal regime. The MRC, charged with sustainably developing the LMRB, has left many inquiries concerning the impacts of the Yali Falls dam unanswered, branding this lauded joint river basin institution with the image of primarily promoting hydropower development.

4.1.1 *The Sesan River*

Rising in the Central Highlands, the Sesan River is one of the largest, and perhaps the most important, left bank tributaries of the middle Mekong. The Sesan flows southwesterly through the Kontum and Gia Lai provinces of Vietnam into Cambodia where it crosses Ratanakiri and Steung Treng provinces before converging with the mainstream at Stung Treng town. This major tributary of the Mekong drains an area of more than 18,700 km^2, with nearly 60 percent (11,220 km^2) of its watershed located in Vietnam and some 40 percent (7,480 km^2) in Cambodia, forming one of the major sub-basins of the LMRB.[803] The length of the Sesan is about 462 km, the greater length (252 km) of which flows through Cambodia.[804] In this northeastern region of the LMRB, which occupies the left bank of the Mekong mainstream from Stung

801 Pearce, *supra* note 373, at 101 (2006) (emphasis added).

802 Philip Hirsch and Andrew Wyatt, "Negotiating Local Livelihoods: Scales of Conflict in the Se San River Basin" (2004) 45 *Asia Pacific Viewpoint* 51, 64–66.

803 MRC, *supra* note 137; Esa Puustjarvi, "Review of Policies and Institutions Related to Management of Upper Watershed Catchments" (Report of Greater Mekong Subregion Watershed Project (Phase II): Poverty Reduction and Environmental Management in Remote Greater Mekong Sub-region, 2004); SWECO Groner (in association with Norwegian Institute for Water Research, ENVIRO-DEV and ENS Consult), "Environmental Impact Assessment on the Cambodian part of the Se San River due to Hydropower Development in Vietnam Electricity of Viet Nam (EVN), Power Engineering Consulting Company No. 1" (December 2006) 21.

804 Hirsch and Wyatt, *supra* note 802, at 53.

Treng to Kampong Cham, there are three major tributaries: the so-called "3S Rivers"—the Sekong flowing from the Lao PDR and Cambodia, and the Sesan and the Srepok rising in Vietnam and traversing westward into Cambodia. The Kontum Highland area, near Kontum and Plei Ku, is the waterhead of the Sesan, while the Darlac Plain, near Ban Me Thuot, is the Srepok waterhead. The Sekong and Srepok watersheds cover about 25,600 km^2 and 30,800 km^2, respectively.[805] The Sekong and Srepok conjoin with the Sesan River in Cambodia before its confluence with the Mekong.

The Sesan proper forms 25 km southwest of the Vietnamese city of Kontum at the confluence of the Dak Bla and Krong Poko rivers.[806] In the 1961 Reconnaissance Report and prior to the Yali Falls Dam construction, the Sesan River was described as follows: 16 km downstream, the river flows through a flat basin surrounded by gently sloping mountains that gradually rise westward closing the valley in steep precipitous slopes.[807] The Sesan then races through a sinuous canyon forming rapids and waterfalls.[808] At the canyon entrance, the river cascades 40–60 m over Yali Falls, the largest waterfall on the Sesan.[809] Farther downstream, still in mountainous terrain, the valley widens and rapids decrease as the river intersects the Cambodian border 55 km from Yali Falls.[810]

The Sesan watershed represents 10.1 percent of the total LMRB. Annual precipitation in the upper Sesan basin is 2,000 to 2,500 millimeters (mm) in its upper reaches of Vietnam and 1,500 mm downstream in Cambodia.[811] At Yali Falls, the drainage area is about 8,300 km^2 and runoff of 40 m^3/sec is available year round.[812] The Sesan contributes about 20.5 percent of the average yearly tributary flow of the average annual Mekong discharge into the delta.[813] The significance of this flow contribution cannot be overstated (see Figure 1.3).

The Sesan is the lifeblood of 50,000 people living along the river in Cambodia, who are not members of the Khmer ethnic group that controls the national government and makes up 90 percent of that country's general population.[814] Ethnic groups in Ratanakiri Province include Brou, Chinese, Javai,

805 1961 Reconnaissance Report, *supra* note 274, at pp. III-26–III-27.

806 V.B. Vladimirov and S.P. Paremud, "Yali Hydro Development on the Se San River in Vietnam" (1997) 31 *Hydrotechnical Construction* 39.

807 *Id.*

808 *Id.*

809 *Id.*; 1961 Reconnaissance Report, *supra* note 274, at p. IV-8.

810 Vladimirov and Paremud, *supra* note 806, at 39.

811 1961 Reconnaissance Report, *supra* note 274, at pp. III-27 and III-38.

812 *Id.* at p. IV-8.

813 Pantulu, *supra* note 170, at 709.

814 Down River, *supra* note 16, at 13.

Kachok, Kavet, Kreung, Lao, and Tampuon.[815] Traditionally, these ethnicities have depended on the river's ecosystem services for their water, food, cultural richness, and livelihoods. For generations the Sesan River provided sustenance to the people living along its banks. However, life on the Sesan changed dramatically following Vietnam's development of the river for hydropower.

4.1.2 *Development of the Yali Falls Dam*

From the earliest days of the Mekong Committee, the Sesan River was viewed covetously by hydrologists and engineers. In the late 1950s, the Mekong and its tributaries were included in a work program undertaken by ECAFE[816] to investigate and study flood control, hydropower, navigation and irrigation in the LMRB.[817] As a result of field studies in 1956, a resolution was adopted at ECAFE's 13th session to promote water resources development in the Mekong.[818] Subsequent to the adoption of this resolution, the four riparian countries of the LMRB—Cambodia, the Lao PDR, Thailand and Vietnam—formed the Mekong Committee with the support of the UN and immediately undertook a Mekong basin investigation.[819] In December 1958, this investigation spawned a report by the UN Technical Assistance Administration entitled Program of Studies and Investigations for Comprehensive Development—Lower Mekong Basin (the so-called "Wheeler Report"), which recommended five major programs.[820] One of the programs recommended in the Wheeler Report was a reconnaissance of major Mekong tributaries.[821] The Japanese Government supported the first reconnaissance survey of Mekong tributaries after the Mekong Committee accepted their proposal at its special session in 1958.[822] Feasibility studies of the hydropower potential of the Sesan River were conducted in the mid

815 *Id.*; Dangerous Waters, *supra* note 16, at 7.

816 ECAFE was renamed the Economic and Social Commission for Asia and Pacific (ESCAP) in 1974.

817 1961 Reconnaissance Report, *supra* note 274, at 5.

818 *Id.*

819 *Id.*

820 *Id.* The UN Technical Assistance Administration was at the time headed by General R.A. Wheeler. The programs of fundamental investigation mentioned in the Wheeler Report consisted of the following tasks: (1) Reconnaissance of major Mekong tributaries; (2) Hydrologic observation; (3) Surveying and mapping (leveling, aerial mapping, supplementary spot leveling, detailed topographic and hydrographic survey, flowage damage survey, construction material survey, geological investigations and borings and hydrographic survey); (4) Related and special studies; and (5) Preliminary planning.

821 *Id.* at 6.

822 *Id.*

to late 1960s by the Vietnamese government in concert with Soviet, Japanese, French, Norwegian, and Swiss firms.[823] The upper Sesan was identified in the 1961 Reconnaissance Report as one of the top four major tributaries in the LMRB that warranted development.[824] As early as 1961, Yali Falls was targeted for its hydropower potential.[825] The 1961 Reconnaissance Report summarized that potential:

> Yali [F]all[s] on the Se San river drop[s] about 60 meters and at this point with the drainage area [of] about 8,300 square kilometers [and] about 40 cubic meters per second of run-off is available even in the dry season. Therefore, utilizing [this] water head and discharge, 20,000 kilowatts of firm electric power could be developed. Besides, when several reservoirs were realized upstream, the firm discharge would be increased to 160 cubic meters per second and ultimately, with the available water head [of] 250 meters, by leading the water through the tunnel of about 7 kilometers, the power development would reach about 400,000 kilowatts in installed capacity and 320,000 kilowatts in firm power output.
>
> As the first stage [of] development, a power generation [of] 30,000 to 40,000 kilowatts utilizing the falls may be sufficient for the purpose of irrigation pumping, general use in Kontum and Plei Ku, and rural electrification. The first stage work would cost approximately 10,000,000 U.S. dollars and the power rate could be fixed very cheap as 0.3 U.S. cent per kilowatt-hour at the receiving end.[826]

The hydropower potential in the Sesan watershed is on its mainstream and, to some extent, on its two main tributaries, the Krong Poko and the Dak Bla Rivers.[827] Additional potential for high-head developments exists on smaller tributaries.[828] In 1993, the Vietnamese government, under the auspices of Electricity of Vietnam (EVN), began constructing the Yali Falls dam on the Sesan

823 Vladimirov and Paremud, *supra* note 806, at 39.

824 1961 Reconnaissance Report, *supra* note 274 at p. S-2. The other three major tributaries in the top four were the Nam Ngum in the Lao PDR, the Stung Battambang in Cambodia and the Nam Pong in Thailand. Other tributaries with high potential for development were the Prek Thnot, Stung Sen and Stung Pursat rivers in Cambodia; the Se Done, Nam Lik, Nam Theun, Se Bang Fai and Se Bang Hiang rivers in the Lao PDR; the Nam Gam, Lam Dom Noi and Chaya Poum rivers in Thailand; and the upper Srepok in Vietnam.

825 *Id.* at pp. IV-8 to IV-9.

826 *Id.*

827 SWECO Groner, *supra* note 803, at 18.

828 *Id.*

with 720 megawatts (MW) capacity. By 1997, six dams were under consideration for the river.[829] There are now seven dams on the Sesan Cascade in Vietnam, including the Sesan 4a Dam. Construction began on the Sesan 4a Dam in 2004, and it entered operation in 2011.

Located some 80 km upstream from the Cambodia border, Yali Falls Dam was built at an estimated cost of US $1 billion, which was supplied by foreign donors—notably the Ukraine and Russia.[830] Sweden and Switzerland provided technical support for the project.[831] A 500-km transmission line from the dam to Ho Chi Minh City was built with a US $575 million loan from the World Bank.[832] Completed in 2001, the Yali Falls dam created a 1,037 million m^3 reservoir with an area of 65.5 km^2, from which water is released for hydropower generation.[833] The Yali Falls project features a 70-m high rock fill dam with the spillway located on the left bank.[834] The power facilities on the right bank of the Sesan consist of an intake structure with two headrace tunnels and an underground power station with four 180-MW generating units.[835] Turbine flow is conveyed back to the river via tailrace tunnels.[836] Mean annual flow is 259 m^3/sec and the turbine design discharge is 420 m^3/sec.[837] About 8,500 people were displaced in Vietnam by the construction of the dam and reservoir.[838] It is Vietnam's first and largest hydropower project on the Sesan River and the country's second largest dam.[839]

A limited EIA for the Yali Falls project indicated the impact area of the dam was no farther than 8 km beyond the project site.[840] Environmental effects of construction and operation of the dam beyond 8 km downstream were underestimated or not assessed[841] by EVN, allegedly due to "security concerns" in

829 Vladimirov and Paremud, *supra* note 806, at 39.

830 Down River, *supra* note 16, at 14.

831 *Id.*

832 Ame Trandem, "A Vietnamese/Cambodian Transboundary Dialogue: Impacts of dams on the Se San River" (2008) 51 *Development* 108.

833 Chris Lang, "The Legacy of Savage Development: Colinisation of Vietnam's Central Highlands" (November 1995–February 1996) 1(2) *Watershed* 43.

834 SWECO Groner, *supra* note 803, at 38.

835 *Id.*

836 *Id.*

837 *Id.* at 40.

838 IRN, *Sesan River*, available at <http://www.internationalrivers.org/campaigns/sesan-dams> (accessed 21 June 2017).

839 Trandem, *supra* note 832, at 108; Lang, *supra* note 833, at 43.

840 Se San Protection Network (SPN), "Learning from Transboundary Environmental Conflicts" (October-December 2003) 6(4) *Mekong Update and Dialogue* 5.

841 Hirsch and Wyatt, *supra* note 802, at 54.

Cambodia.[842] There was no official EIA conducted to determine downstream impacts in Cambodia by EVN, the Cambodian national government, or the MRC. This woefully inadequate EIA has been the subject of intense criticism.[843]

By 1996, construction and operation of the Yali Falls dam had altered the flow regime of the Sesan and adversely affected the downstream villages and communities in Cambodia.[844] Massive flooding occurred that year along the river in Cambodia, ostensibly resulting from the failure of the cofferdam (a temporary dam built to divert water around the dam site to allow construction).[845] By the time the dam was operational in 2002, large-scale environmental, social, and economic impacts had occurred, including modification of the river's hydrology and water quality.[846] Approximately 50,000 people in downstream communities were impacted.[847] Between 1996 and 2003, the Sesan's hydrological regime and biodiversity were severely altered as evidenced by irregular and frequent flooding, unusual flood pulse events, dry-ups, increased dry season flows, daily flow and height fluctuations, water quality problems, and fish and fisheries declines.[848]

Construction and operation of the dam caused flood pulses along the entire length of the river. Oxfam attorney Michael Lerner documented 39 deaths by drowning that were attributable to unpredictable and rapid changes in river flow and height caused by wet season flood control operations at Yali Falls.[849] Flood pulses from the dam's operation drowned livestock, swept away fishing equipment and boats and devastated riverbank agriculture.[850] Water level rises also occurred in the dry season, disrupting fishing operations and agricultural practices along the river banks. The MRC recorded river level and flow fluctuations occurring on a diurnal basis and at other unpredictable periods separated by days.[851] These year-round erratic flows and height fluctuations, unrelated to rainfall events, are attributed to dam operations during hydropower production when water is passing through the dam's turbines.[852] In the rainy season, rapid water level rises are most likely caused by spillway releases

842 Down River, *supra* note 16, at 14.

843 SPN, *supra* note 840, at 5.

844 Hirsch and Wyatt, *supra* note 802, at 54; Dangerous Waters, *supra* note 16, at 4.

845 Trandem, *supra* note 832, at 109.

846 Hirsch and Wyatt, *supra* note 802, at 54; Trandem, *supra* note 832, at 108.

847 Hirsch and Wyatt, *supra* note 802, at 56.

848 *Id.* at 56–58; Trandem, *supra* note 832, at 109.

849 Michael Lerner, "Draft Se San Legal and Policy Study" (Oxfam America, 2002).

850 Hirsch and Wyatt, *supra* note 802, at 57; Trandem, *supra* note 832, at 109.

851 Hirsch and Wyatt, *supra* note 802, at 57.

852 *Id.*

for the Yali Falls reservoir.[853] Dry season water fluctuations are most likely resultant from the Yali hydropower plant operations.[854] Once an important alternative source of income to villagers, gold panning became dangerous due to these erratic pulses.[855]

Increased turbidity in the river is attributed to increased river bank erosion and scouring associated with the river level fluctuations. Downstream populations on the Sesan have customarily used river water for domestic uses, especially drinking and bathing. Since 1996, villages have reported increases in illnesses including stomach ailments such as diarrhea and vomiting after drinking from the river, as well as skin rashes and eye irritations after bathing in the river.[856] In addition, 952 deaths have been attributed to changes in water quality and thousands of animals have perished after drinking river water.[857] After dam construction, breakdown of terrestrial vegetation can occur in the reservoir, releasing nutrients into the deep water which may produce toxic blue green algae. A water quality study conducted in 2005 at the Yali Falls reservoir found low concentrations of algal toxins and the algal species *Microcystis* known to produce these toxins.[858] The symptoms, deaths, and losses of animal life along the Sesan are consistent with impacts from blue green algal toxins and may have resulted from development of algae in the river downstream due to release of nutrient rich bottom water from the Yali Falls reservoir.[859] A 2009 technical report confirmed high levels of toxic algae and bacteria in the Sesan River, which exceeded World Health Organisation limits for safe drinking water.[860] Interestingly, the bacteria were not found in three Sesan tributaries, suggesting that this pollution originated from a more stagnant source such as an

853 SWECO Groner, *supra* note 803, at 11.

854 *Id.* at 11–12.

855 Trandem, *supra* note 832, at 109.

856 Hirsch and Wyatt, *supra* note 802, at 58; Trandem, *supra* note 832, at 109.

857 World Rainforest Movement, *Dam in Vietnam hits Cambodians* (January 2001) Bulletin No. 42 at 1, available at <http://wrm.org.uy/oldsite/bulletin/42/Vietnam.html> (accessed 16 June 2017).

858 SWECO Groner, *supra* note 803, at 129.

859 *Id.* at 128–131.

860 *See* Anna Madeleine Tiodolf, *A Limnological Study in the Se San River in Cambodia During the Dry Season: Focus on Toxic Cyanobacteria and Coliform Bacteria* (unpublished Master thesis, Norwegian University of Life Sciences, 2009); Anna Madeleine Tiodolf and Per Stålnacke, "A Limnological Study in the Se San River in Cambodia During the Dry Season: Focus on Toxic Cyanobacteria and Coliform Bacteria" (STRIVER Technical Brief No. 12, 2009) 4.

upstream reservoir.[861] This study indicates that cyanobacteria and cyanotoxins in the Sesan River proper are linked to hydropower reservoirs.[862] According to the 3S Rivers Protection Network (3SPN), "These results are consistent with the gastric disorders and skin problems experienced by Sesan River communities since the construction of the Yali Falls dam."[863] These research findings seem particularly relevant given Cambodia's plans to construct five to seven dams on the Sesan and its tributaries, including the 400-MW Sesan II hydropower facility in Stung Treng, which will resettle more than 5,000 people, inundate over 33,000 hectares (ha) of land and further exacerbate fish and fisheries problems.[864] The Lower Sesan 2 Dam will be completed by 2019. The O Chum 2 Dam was built on a Sesan tributary in 1992.[865]

Fish catches have declined along the Sesan. Fish are the main source of protein and an important source of income for local populations. Results of impact studies on river fisheries indicate that the dam has "generally led to serious negative downstream impacts on fish, fisheries and other aquatic life in the Sesan River in Ratanakiri province, [in] northeast Cambodia."[866] Increased turbidity, high sediment loads, alteration of the algae content in the water, and non-seasonal fluctuations, all resulting from dam construction and operation, may explain the reasons for the fish and fisheries declines of 10–30 percent over the past decade.[867]

Deep pools on the Sesan play a crucial role in the river's ecosystem because they provide habitats for many fishes in the dry season. These "deep pool habitats have been affected by siltation as a result of changed hydrological conditions" because of construction of the Yali Falls dam.[868] One deep pool in

---

861 Mouth to Source, *3SPN-Dams on Sesan River Polluting Water and Poisoning Downstream Communities* (8 September 2009) <http://khmerization.blogspot.com/2009/09/new-scientific-study-dams-on-sesan.html> (accessed 16 June 2017).

862 *Id.*

863 *Id.*

864 *Id.*

865 CWPF-Mekong, *Dams of Cambodia in the Mekong Basin* (2013) <http://wle-mekong.cgiar.org/download/map-repository/A4_Cambodia_Element.pdf> (accessed 16 June 2017).

866 Ian G. Baird and Meach Mean, *Sesan River Fisheries Monitoring in Ratanakiri Province Northeast Cambodia: Before and After the Construction of the Yali Falls Dam in the Central Highlands of Vietnam*, 3S Rivers Protection Network (3SPN) and Global Association for People and the Environment (GAPE) (2005) 14.

867 Hirsch and Wyatt, *supra* note 802, at 58; SWECO Groner, *supra* note 803, at 13.

868 MRC, Deep pools as dry season fish habitats in the Mekong River Basin, Technical Paper No. 4, at 3 (9 April 2002).

Voen Say district "diminished from 7–8 meters to just half a meter in depth."[869] Consequently, many fish species in the district have declined or almost completely disappeared.[870] A preliminary EIA report published by "Swedish Consultants" (SWECO) Groner on behalf of EVN in 2006 concluded that "it is *very likely* that the Yali [Falls dam] regulation has contributed significantly to the observed decline in fish."[871] Impacts on fish and fisheries threaten food supplies and livelihood security along the Sesan River.[872]

Australian consultants for the Asian Development Bank, Worley WTL, Ltd., described the dangerous conditions in the Sesan River in an April 2000 report, now known as the Worley Report.[873] This report singled out EVN as the responsible party for damages associated with construction and operation of the Yali Falls dam.[874] The Worley Report stated:

> Over the centuries a knowledge of river flows has been built up which has informed villages about safe practices and strategies for resource use, including recreation use. High floods in the dry season and rapid rates of change in river level are so rare that when they do occur they find villagers "unprepared" .... In some cases ... people have been at the wrong place at the wrong time, and have drowned.
>
> Two of the spills [in 1999 were] over 4000 m^3/s. A flow of 4000 m^3/s in a channel with a gradient of 0.5% is not survivable for people in the water. Even 500 m^3/s in such a channel will probably kill all children and half the adults who are caught in the water.
>
> What happens when a spill of this magnitude is initiated in the dry season is this: a river bed is transformed unexpectedly and quickly (in a few minutes) from an area which is safe to walk in, interesting to play in the shallow pools, productive to fish, generally crossed by wading and dotted with islands where people set up dry season fishing camps, into a lethal turbulent channel in which people *cannot survive*. Deep undertows can trap even strong swimmers in underwater circulation systems for many minutes. People swept down rocky chutes suffer numerous violent

869 *Id.* at 11.

870 *Id.*

871 SWECO Groner, *supra* note 803, at 13 (emphasis added).

872 Trandem, *supra* note 832, at 109.

873 Down River, *supra* note 16, at 23; Worley WTL, Ltd., "Se San 3 Hydropower Project: Yali/Se San 3 Environmental and Social Impact Analysis Study" (Asian Development Bank-PPTA 31362-01-V16, RMF Draft Copy, April 2000) 2-22 to 2-23 [hereinafter Worley Report].

874 Worley Report, *supra* note 873, at 2-23; Down River, *supra* note 16, at 87, endnote 6.

> collisions with rocks which eventually kill them, or leave them too damaged to swim, so that they drown even in the relatively calm waters downstream of the chute.[875]

In response to the crises attributed to Yali Falls dam, a village representative network, the Sesan Protection Network (SPN), was established to address the difficulties faced by the river communities and to document the environmental, social and economic impacts.[876] The SPN was formally launched in December 2001 and by 2004 included all 60 villages in Ratanakiri Province in Cambodia.[877] The SPN established a multifaceted advocacy coalition including non-governmental organizations (NGOs), allies within civil society, local government authorities, and the media, which used strategies such as lobbying, public meetings, letter writing, and media outreach to create political attention and generate international pressure.[878] SPN has interacted with the MRC, requesting that they establish dialogue and communication among all the stakeholders in order to address the effects of the Yali Falls Dam and asking them to implement basin-wide transboundary water governance and conflict management by enforcing the 1995 Mekong Agreement.

On 24 November 2002, the ethnic minorities in Ratanakiri Province issued a Joint Statement appealing to the national government of Cambodia to provide assistance.[879] This statement decried the greatest impacts of the construction and operation of Yali Falls dam:

- villagers' safety—a number of villagers have died due to the floodings and more have been affected by disease;
- villagers' food security—villagers are facing increased food shortages every year as their ability to cultivate along the river is diminished;
- water levels and quality—unusual fluctuating water levels have disrupted farming and fishing practices and a degeneration of the water quality has caused various skin diseases and ailments;
- fish—fish species and stocks have declined as deep water pools become shallow;

---

875 Worley Report, *supra* note 873, at 2-22 to 2-23; Down River *supra* note 16, at 87, endnote 6 (emphasis in original).

876 Trandem, *supra* note 832, at 109.

877 *Id.*

878 *Id.* at 110.

879 SPN, Ratanakiri Se San Communities, "Joint Statement from Ethnic Minority Groups Living Along the Sesan River in Ratanakiri" (Ratanakiri Province, Cambodia, 2002).

- farming, rice production, and other cultivation which have all suffered due to water fluctuations and flooding;
- villagers' houses and farm houses have been destroyed;
- domestic and farm animals which have died or been lost;
- fishing equipment and boats that have been swept away or destroyed;
- riverine animal and bird species [which] have died;
- livestock which have declined in numbers due to the death of cattle and buffalo and increases in cattle diseases;
- fishing which has declined as people are afraid to sleep on the river islands for fear of swift floodings;
- river depth—the river has become shallow due to erosion and sand deposits;
- gold panning activity (gold panning holes filled by sand deposited by the river) that has declined;
- wild vegetables along the river that have been uprooted and destroyed;
- river banks which have suffered erosion and a loss of trees;
- traditional culture—destruction to livelihoods and ways of living have adversely affected time-honored solidarity amongst communities and inter-village relationships;
- vegetable production in the dry season which has been lost due to floods;
- forests that are being lost as villagers are forced to abandon their rice fields and look to the forests for food sustenance and income; [and] infrastructure along the river such as roads, bridges, schools, and administrative buildings which has been destroyed.[880]

Collectively, the villagers requested that: (1) the government help stop the continued construction of hydropower dams on the Sesan River; (2) the natural flow of the river be restored; (3) compensation for damages be paid; (4) the government of Cambodia negotiate with the government of Vietnam to resolve the dam's impacts; and (5) the MRC and stakeholders conduct further studies in consultation with the people along the river.[881] As a result of this advocacy, Cambodia and Vietnam finally acknowledged the negative downstream impacts caused by the Yali Falls dam. In November 2002, a representative of Vietnam's Ministry of Industry stated "we are very sorry for the losses of the people living downstream on the Se San River caused, of course, by releasing water [without prior notification] from Yali Falls dam's reservoir in February 2002."[882] Later in August 2003, the Cambodian NMC sent a facsimile to Vietnam's NMC

880 *Id.* at 1. (bullets added for emphasis).

881 *Id.* at 2.

882 SPN, *supra* note 840, at 6; Trandem, *supra* note 832, at 111–112.

requesting a response to the Sesan communities' concerns.[883] Public opposition to further dam construction on the river led the Asian Development Bank to request additional EIAs to fully address the Yali Falls dam's downstream impacts.[884] Vietnam promptly alerted the Bank that it "no longer needed [its] assistance to proceed with the project."[885] In December 2006, SWECO Groner on behalf of EVN, conducted a full EIA of the effects of hydropower development on the Sesan which documented many of the negative impacts of the Yali Falls dam on the river downstream in Cambodia.[886]

The impacts from the Yali Falls Dam—unusual flooding events, increased dry season flows, unpredictable fluctuations in river flow and height, degraded water quality, health impacts, and fish and fisheries declines—have had devastating human consequences. Damages from these impacts were estimated to exceed US $3.3 million.[887] Additionally, the economic costs may not be outweighed by the benefits provided from hydropower development in the Mekong power grid.[888] Perhaps most disconcerting, from the outset of the Yali Falls project, more than 50,000 people in downstream communities were not notified nor consulted during construction and this has remained the case throughout the operation of the dam.[889] More than 15 years after the dam's official commissioning in 2002, environmental, social, and economic impacts have yet to be remedied or resolved,[890] raising serious questions about

---

883 Trandem, *supra* note 832, at 112.

884 *Id.*

885 Wayne C. White, "A Review of the Se San 3 Hydropower Feasibility Study" (prepared for Probe International, 2000).

886 SWECO Groner, *supra* note 803, at 127–152.

887 Down River, *supra* note 16, at 53; Andrew B. Wyatt and Ian G. Baird, "Transboundary Impact Assessment in the Sesan River Basin: The Case of the Yali Falls Dam" (2007) 23 *Int'l J. of Water Resources Development* 427, 433.

888 IRN, "Sizing up the Grid: How the Mekong Power Grid Compares Against the Policies of the Asian Development Bank" (2004) 10. [hereinafter IRN (2004)] The "Mekong power grid" is the grandiose Asian Development Bank initiative "promoting the development of a regional power grid and electricity trading system in Southeast Asia fueled by hydropower. The Asian Development Bank envisions that China's Yunnan province, Burma and Laos-where hydropower potential is huge and community opposition is stifled-will generate cheap, reliable and environmentally sustainable power for growing markets in Thailand and Vietnam. Hydropower projects are also recommended for development on Cambodia's [section] of the Se San River ... and rivers in central and southern Laos, among others."

889 Osborne (2004), *supra* note 208, at 32–33; Trandem, *supra* note 832, at 109.

890 Trandem, *supra* note 832, at 109.

the efficacy of the sustainable development paradigm enshrined in the 1995 Mekong Agreement.

Yali Falls was the first litmus test for the MRC. The MRC informed Cambodia and Vietnam that it was "contrary to the spirit of the 1995 Mekong Agreement to have natural resources developed in one country at the expense of people in another country," and recommended "compensation for people already affected by the operation of the current dams."[891] Initially, the only inquiry of the MRC into the events surrounding the Yali Falls dam disaster was a four-day research investigation to Ratanakiri Province in 2000 followed by facilitating the establishment of the Cambodia-Vietnam Joint Committee for the Management of the Sesan River.[892] This Joint Committee's main agenda item was procedures for advance notices of water releases which were established but remain ineffective and problematic. Neither the MRC nor the Cambodia-Vietnam Joint Committee for the Management of the Sesan River (now the "Standing Committee") has dealt with compensation issues or provincial requests for a moratorium on dam construction on the Sesan.[893] The CEO of the MRC stated in 2002 that the Yali Falls matter was the responsibility of the Cambodian Government and outside the jurisdictional mandate of the MRC.[894] The Cambodian NMC requested that the MRC Environment Programme undertake a study of the water quality of the Sesan River.[895] Not surprisingly, the results of this study indicated water quality was still good, although conspicuously missing were tests for pathogens and filamentous algae.[896] On 7 July 2007, Vietnam publicly pledged to preserve the water resources and ecosystem of the Sesan River Basin.[897] Vietnam's temporary closing of the Yali Falls dam on 2 February 2010 caused the Sesan River in Northeast Cambodia to become so dry that is was possible to walk from one side to the other.[898] Water fluctuations from the dam closings continue to impact fish and fisheries.[899] A spokesman for the

891 MRC, Report Regarding Terms of Reference for Studies on Se San River Water Management (2001).

892 Wyatt and Baird, *supra* note 887, at 438; The Joint Committee for the Management of the Se San River was reconstituted as a "Standing Committee" later in 2000.

893 Trandem, *supra* note 832, at 112.

894 Hirsch and Wyatt, *supra* note 802, at 62.

895 MRC, Annual Report 2006 at 17.

896 *Id.*

897 Viet Nam News, *Viet Nam Vows to Preserve Water Resources, Eco-system* (7 July 2007).

898 Cameron Wells and Tha Piseth, *Closed Dams a Headache for Fishermen* (11 February 2010) The Phnom Penh Post < http://www.phnompenhpost.com/national/closed-dams-headache-fishermen> (accessed 18 June 2017).

899 *Id.*

3SPN stated in an interview in 2010 that "the recent temporary closures of the Yali Falls dam had not only threatened fishermen's livelihoods, but had also made some of them ill."[900]

To this day, reparations have not been paid by the Vietnamese Government to the Cambodian Government, let alone the villagers whose livelihoods have been transformed by the Yali Falls Dam. Moreover, many of the same villagers are now at risk at being forcibly removed from their ancestral homelands; essentially being squeezed out by hydropower projects from both upstream and downstream. The villagers of Sre Ko, for instance, were left out of the stakeholder consultation and outreach process during the planning phase of Lower Sesan 2, presumably because they are members of an ethnic minority.[901] The villagers have been offered compensation for their land and the value of their crops, but many plan to remain and some are even planning to construct floating houses from bamboo, having no intention to leave once the area is flooded by Lower Sesan 2.[902] Even for locals who will not be directly displaced by the creation of the reservoir, an increased risk of death and destruction of property looms due to the persisting occurrence of flash floods from Yali Falls dam. Many locals maintain that the warning system for major releases from Yali Falls is ineffective and unreliable, and they feel that their voices have ultimately been ignored by the Cambodian government and MRC feasibility studies.[903] These concerns are compounded by the latest dam being built. As one villager reflected, "The flash floods [from releases at Yali Falls] destroy crops and sweep away livestock as well as people. We are caught in the middle. If they finish Lower Sesan 2, how can we survive when Yali opens its reservoir and the Lower Sesan 2 blocks the water in its reservoir?"[904] Compounding the complexity of the situation regarding the Lower Sesan 2 project is the opposition from Vietnam. These concerns stem from stakeholders in the Mekong delta in southern Vietnam, who are wary of the predictions that Lower Sesan 2 will obstruct a significant amount of sediment flowing in the Sesan, ultimately impairing the vitality of the rich, but fragile soil and nutrient systems in the delta.[905] It would appear that, in this case at least, the shoe is on the other foot, since Vietnam

900 *Id.*

901 May Titthara, *Trapped between Two Dams* (7 January 2016) Khmer Times <http://www.khmertimeskh.com/news/19609/trapped-between-two-dams/> (accessed 18 June 2017).

902 *Id.*

903 *Id.*

904 *Id.*

905 Rod Harbinson, *Cambodian Dam Proceeds Despite Opposition over Fish, Ousted Villagers* (17 August 2015) Mongabay <https://news.mongabay.com/2015/08/cambodian-dam-proceeds-despite-opposition-over-effects-on-fish-ousted-villagers/> (accessed 20 June 2017).

failed to consider the well-being of Cambodian villagers who depended on the free-flowing Sesan when constructing the Yali Falls project—let alone its own delta farmers.

In response to the multifaceted criticisms, Stung Treng provincial officials (where Lower Sesan 2 is located) have countered that the Lower Sesan 2 project was approved by the Cambodian National Assembly after an adequate environmental review process, and point out the path toward economic development that will be realized by the project's completion; namely that four Cambodian provinces will no longer have to import electricity from the Lao PDR.[906] Until recently, the MRC maintained that the ongoing issues affiliated with the 3S Rivers Basin to be under the jurisdiction of the Cambodian and Vietnamese governments, and therefore outside of their purview. However, on 24 June 2015, the MRC released a statement expressing the view that the Prior Constellation process should apply to hydropower projects on Mekong tributaries that might impact neighboring countries.[907] Nonetheless, this does not apply retroactively to the Cambodian government in the case of Lower Sesan 2, and there is little evidence that the concerns of local groups and stakeholders in Vietnam and other nations are being adequately considered by developers.[908] The ongoing disenfranchisement of these vulnerable populations that rely on Mekong tributaries to sustain themselves, compounded by the apparent inability of the national governments to work together in addressing these issues, further highlights the marginalization of tributaries as a central failure of the status quo Mekong legal regime.

With several more dams planned on the Sesan, concerns are heightened about water quality, which is inextricably linked to water quantity, and riverbank erosion.[909] The aftermath of the Yali Falls debacle unveiled the need for the MRC, NMCs, and various other actors to promote compliance with international law.

4.2 The Thai Water Grid

In 2003, the Thai Government launched a US $5 billion grandiose plan to divert water to unirrigated farms and to Bangkok to alleviate the effects of "natural allocation."[910] Officially named the "Sustainably Holistic Water Management

906 Titthara, *supra* note 901.

907 Harbison, *supra* note 905.

908 *Id.*

909 *Id.*

910 *See generally,* Larry O. Putt, "Water Resources Protection in Alabama: The Need for a Paradigm Change" (2003) 7 *T.G. Jones L. Rev.* 1, 7–8. (Often "the areas of greatest demand

Project," this plan involved the construction of a network of pipes across Thailand to increase irrigated land by fourfold and to ensure the population's access to potable water. The plan, designed to accommodate Bangkok's burgeoning population and to irrigate the dry northeastern region of Thailand, or "Isan" where agriculture is one of the main activities, includes significant transboundary water diversions from 18 sites including important Mekong tributaries such as the Stung Num River in Cambodia and the Se Bang Fai, Se Bang Hiang, and Nam Ngum rivers in the Lao PDR.[911] Borrowing from the power sector, the project was dubbed the "Thai water grid" to describe the interconnectivity of its multiplicity of dams, canals, pipes, and natural waterways.[912]

Water diversions from the Stung Num watershed in Cambodia are viewed by the Thai Government as a suitable solution to water shortages in eastern Thailand.[913] Under this plan, some 200 km of pipeline would divert water from a proposed dam on the Stung Num River to the Prasae Reservoir in Rayong Province, Thailand.[914] The Thai Department of Water Resources acknowledges that because of the project's "large scale and high investment cost," the proposed water diversion would take considerable time and additional study before it could materialize.[915] The Stung Num water diversion has been part of the Thai water grid for some time, but was revisited most recently in 2005 because of the water needs of petrochemical plants, which resulted in local water diversions from many rivers and attendant conflict among the competing users in the eastern provinces.[916] In 1992, Cambodia and Thailand agreed to develop a hydropower project which would require water diversions of 546 million m^3 from the Stung Num watershed.[917] According to a feasibility study report, water use in Thailand's eastern provinces will increase from 295 million m^3 in 2007 to about 600 million m^3 in three decades (2037).[918]

The Nam Ngum, Se Bang Fai, and Se Bang Hiang rivers were identified as major tributaries for potential development as early as 1961.[919] The Nam Ngum,

---

on supplies are the areas with the least supplies. This produces a kind of natural allocation. When [t]here are no current legal mechanisms for reallocation ..., it is unlikely that current land use patterns will change radically." (citations omitted)).

911 Hirsch and Jensen, *supra* note 180, at 148.

912 *Id.*; Molle and Floch (2008a), *supra* note 14, at 200.

913 Apinya Wipatayotin, *Cambodia Eyed as New Source of Water Supply* (28 August 2007) Bangkok Post.

914 *Id.*

915 *Id.*

916 *Id.*

917 *Id.*

918 *Id.*

919 1961 Reconnaissance Report, *supra* note 274, at p. S-2.

like the Sesan, was in the top four tributaries selected for development in the 1961 Reconnaissance Report.[920] With a watershed of about 16,000 km^2, including the basin of its tributary the Nam Lik, the Nam Ngum is the site of a major dam from which the Lao PDR has been selling hydropower to Thailand since its completion in the 1970s.[921] This was the first major dam erected on a tributary in the LMRB and it remains highly controversial. The Se Bang Fai is a medium sized tributary of the Mekong located in the central Lao PDR, has a watershed of approximately 9,550 km^2, and conjoins with the mainstream 55 km downstream of Thakhek.[922] Rising in mountainous areas, the Se Bang Hiang forms the border between the Savannakhet Province in the Lao PDR and the Quang Tri Province in Vietnam and flows across the former province until its confluence with the Mekong. It is one of the largest Mekong tributaries in the Lao PDR, with a watershed of some 19,800 km^2 and annual runoff of 16,000 million m^3.[923] Many potential dam sites are found on both the Se Bang Fai and the Se Bang Hiang tributaries, where annual precipitation ranges from 2,000 to 2,500 mm.[924]

For decades the Thai government has been preoccupied with the "desert bloom syndrome,"[925] the concept of expanding irrigation in the poor and dry region of Isan. Aimed at enabling dry season irrigation, two previous plans dubbed "Isan Khiew" ("Green northeast") and the Khong-Chi-Mun water diversion proposed a vision with a simple idea: water would be taken from places where there was too much and reallocated to places where there was too little in an effort to offset natural allocation. The former scheme (in 1987) involved military development projects, and the latter (in 1989) envisioned diversion of Mekong water into the headwaters of the Chi and Mun rivers and construction of a network of irrigation channels.[926] Covering about two-thirds of northeast Thailand, the Chi and Mun river basins form the largest sub-basin of the Mekong

920 *Id.*

921 Osborne (2004), *supra* note 208, at 35.

922 1961 Reconnaissance Report, *supra* note 274, at p. IV-39.

923 *Id.* at p. IV-41; Southeast Asia Rivers Network (SARN), "A Summary Report: Conceptual Study Report on Laos-Thai Friendship Water Development for Sustainable Agriculture in Savannakhet Province of Lao P.D.R. and Lower Chi Basin of Thailand" (conducted by Sanyu Consultants Inc., June 1998; prepared by Southeast Asia Rivers Network-Thailand Chapter, 20 June 2002) 1.

924 1961 Reconnaissance Report, *supra* note 274, at pp. IV-39 and IV-41.

925 *See generally,* Francois Molle and Philippe Floch, "The 'Desert Bloom' Syndrome: Politics, Ideology, and Irrigation Development in the Northeast of Thailand" (IWMI Working Paper 4, 2008b).

926 Hirsch and Jensen, *supra* note 180, at 149.

encompassing a drainage area of 119,570 km^2 or 20 percent of the LMRB.[927] The Isan Khiew project never materialized due to lack of funding. The Khong-Chi-Mun project was partially implemented but eventually shelved, due in part to financial crisis and on partially social and environmental grounds.[928] These proposed water diversions were regarded critically in Cambodia and Vietnam, where concern centered on the extent to which the diversions would reduce flow in those states' reaches of the Mekong.[929]

Although smaller scale options, such as diversions from the Mekong mainstream, are more economically and technically sound, the Thai Government opposes them because they would involve multilateral negotiations with all LMRB states under the terms of the 1995 Mekong Agreement.[930] Emblematic of this position is the following statement in the Thai water grid pre-feasibility document:

> The feasibility of the diversions from neighboring countries gives more weight to considerations relating to agreements with the relevant countries than other considerations such as the investment costs. This is because a number of rivers that are the sources of water for these diversions are international rivers [the Salween and its tributary the Muoi and the Mekong].[931]

Clearly, the perception is that it is easier for the Thai government to target development of Mekong tributaries than attempt to obtain consensus for mainstream diversions under the 1995 Mekong Agreement. Accordingly, in a memorandum of understanding in 2004, the Thai and Lao PDR governments authorized diversions from tributaries in the Lao PDR into Northeast Thailand through pipes *under* the Mekong, or as Dore and Lazarus suggest "siphoning under the 1995 Mekong Agreement,"[932] in a water trade dubbed the "Thai-Lao Water Friendship."[933] This bilateral cooperative agreement for irrigation and agridevelopment would (1) prevent flooding in the Xonburi and Champhon districts of the Savannakhet Province, Lao PDR; (2) provide irrigation water

927 Tawatchai Tingsanchali and Purushottam Raj Singh, "Optimum Water Resources Allocation for Mekong-Chi-Mun Transbasin Irrigation Project, Northeast Thailand" (1996) 21 *Water International* 20.

928 Molle and Floch (2008a), *supra* note 14, at 200.

929 Osborne (2004), *supra* note 208, at 35.

930 Hirsch and Jensen, *supra* note 180, at 149.

931 *Id.*

932 Dore and Lazarus, *supra* note 362, at 362.

933 *Id.*

for farm lands in the main and tributary watersheds of the Se Bang Hiang and the lower Chi Basin of Northeast Thailand; and (3) provide water storage in the Se Bang Hiang reservoir during the dry season for both states.[934] This agreement proposes two dams, the Se Bang Hiang 1 and 2, which would cut flood discharge by 2,000–3,000 m^3/s and store the discharge in the reservoir.[935] Construction of the two Se Bang Hiang dams and water diversion facilities would displace 20–30 villages.[936]

The Thai water grid is a glaring example of a "megaproject": a big idea that makes very little sense.[937] The grid would triple Thailand's irrigated area from 5 million to 17 million ha within five years with an expected benefit to enable farmers to cultivate and access water year-round.[938] Spun by politicians as a plan to "eradicate poverty," the project was aimed primarily at Thailand's dry northeastern region, Isan.[939] However, due to its costs, scale, complexity and serious economic, social and environmental implications, the project met instant opposition from environmentalists, economists, and water management experts.

Diversions of tributary waters in the Lao PDR are bound to have unprecedented environmental impacts on the sub-basins, not to mention the lost opportunities for future benefits for economic development, irrigation, recreation, instream water values, and other water uses. Accordingly, in addition to the costs of the water itself, two elements of sustainable development, *inter*generational and *intra*generational equity, dictate that the Lao PDR should request that Thailand provide a sub-basin of origin equity fund[940] to compensate the areas for water extraction and losses. Additionally, a megaproject such as the water grid will inevitably have significant social effects in both countries. Anticipated impacts in the Lao PDR would include:

- Displacement and resettlement of indigenous peoples in several villages located in the proposed reservoir site;

934 SARN, *supra* note 923, at 1.
935 *Id.* at 2.
936 *Id.* at 5.
937 *See generally,* Molle and Floch (2007), *supra* note 14; Molle and Floch (2008a), *supra* note 14.
938 Molle and Floch (2008a), *supra* note 14, at 200.
939 Molle and Floch (2007), *supra* note 14, at 12; Molle and Floch (2008b), *supra* note 925, at 14–15.
940 Vujica Yevjevich, "Water Diversions and Interbasin Transfers" (September 2001) 26 *Water International* 342, 346.

- Blockage by dam construction of the migration route of fish species that swim upstream from the Mekong to spawn in the Se Bang Hiang tributary;
- Disruption by the dam of both the ecosystem and life cycle of fish in the Mekong and the Se Bang Hiang;
- Inundation of villages and farm land by the proposed reservoir; and
- Construction and operation of the dams could cause unusual and erratic flood pulses, flow regime alteration, river level fluctuations and water quality problems similar to those attendant to Yali Falls.

In Thailand, diverting large amounts of water could:

- Increase hypersaline soil conditions, already a major problem in Isan[941] (e.g., Rasi Salai dam in Si Sa Ket Province has increased soil salinization in farm land in the higher elevations through irrigation by hypersaline water from the dam reservoir in lower elevations); and
- Increase costs of water for local farmers.

The Thai-Lao PDR transboundary water diversions have been justified in the latter country by advancing the economic benefits from "the sale of supposedly excess water seen as a positive for the national economy"[942] without significant input from civil society. By circumventing the 1995 Mekong Agreement through the bilateral Thai-Lao Water Friendship, which the former agreement invites,[943] the parties attempt to justify their positions with reference to national interests. Effectively emulating US Attorney General Harmon in the 1895 Rio Grande dispute, the Thai Director of the Department of Water Resources stated that "it would be a violation of national sovereignty if a nation could not implement development projects or use water from its rivers independently."[944]

One of the most critical environmental problems of Northeastern Thailand is soil hypersalinity.[945] It is difficult to imagine how the large-scale water diversions and irrigation plans of the Thai water grid could do anything but increase the salinity problems in Isan. Indeed, a 1995 report on the Mun Water Resources

941 Francois Molle, "Irrigation and Water Policies in the Mekong Region: Current Discourses and Practices" (IWMI Research Report 95, 2005) 3.

942 Hirsch and Jensen, *supra* note 180, at 150.

943 1995 Mekong Agreement, *supra* note 1, art. 38.

944 Rules being drafted not fair to Thailand. Three-year talks still far from a deal, *Bangkok Post* (11 November 2003).

945 Molle and Floch (2007), *supra* note 14, at 16–17.

Development Master Plan Study observed that "[i]nsufficient data exist on the inter-relationship between surface water, ground water movement, and salinity, and there is a high risk of *increasing salinity hazard* by developing ground water or transferring surface water in the basin on a large scale."[946] An expert committee selected by the Thai Office of Environmental Planning and Policy confirmed that sub-basin and interbasin diversions and irrigation would exacerbate salinity problems.[947]

Massive cultivation in Northeastern Thailand can be expected to cause dry season pollution and human health hazards derivative from extensive use of agrochemicals. Severe environmental impacts would occur as a result of a switch from conventional to intensive agriculture and the attendant use of hazardous chemicals in the fields.[948] These chemicals would also threaten fisheries in the Mekong mainstream and tributaries.[949]

The Thai water grid is a classic example of how megaprojects in Southeast Asia are justified by (1) bringing to a fever pitch water shortages and the effects of flooding; and (2) considering only a megaproject's national benefits while disregarding its costs.[950] Many "pay-for-piped water projects"[951] in Thailand have failed because farmers refused to pay for the "cost of electricity and associated operation and maintenance fees required to pump water onto their land."[952] Hirsch and Jensen observed that:

> Irrigation officials have recommended that farmers shift from rice-cultivation to more profitable cash crops and pay fees for water use to cover the costs of maintaining pipelines. This has been an unpopular proposal in a context where water fees and pay-per-use models do not mesh well with the past expectation that government financed irrigation schemes provide water for free. Moreover, in the absence of metering the proposed payment is not based on volumetric consumption, so

946 Binnie and Partners, "Mun River Basin Water Resources Development Master Plan: Final Technical Report" (1995).

947 I. Wisznicwski, "Risking Salinity in Thailand and Lao PDR" (2003) 9(1) *Watershed* 12–25.

948 Bangkok Post, *Water crisis looms, says grid study* (13 June 2004).

949 L. Kamkongsak and M. Law, "Laying Waste to the Land: Thailand's Khon-Chi-Mun Irrigation Project" (2001) 6(3) *Watershed* 25.

950 The focus on national benefits rather than on cost-benefit analysis is summarily provided by a quote from a high-ranking Thai official who stated that the Thai water grid was "a worthwhile investment because it will benefit 30 to 40 million people nationwide." *The Nation* (23 June 2006).

951 Hirsch and Jensen, *supra* note 180, at 149.

952 *Id.*

> the change is seen as more of a tax than a user-pays facility or incentive to economize on water use. The expense associated with the project has also been criticized by government officials, who admit the scheme needs to be reduced in size and extent to make it economically and technically achievable.[953]

The Thai water grid, like the Yali Falls Dam project, is emblematic of a common governance trend in the LMRB. Although the grid would have significant environmental, social and economic impacts on large populations, there has been little stakeholder involvement. Indeed, the grid planning process is "shrouded in secrecy."[954] According to the International Water Management Institute, this is "[o]ne common thread that links all the Mekong countries."[955] Noteworthy is the disconnect between public participation and megaprojects:

> It's also clear that the principles of stakeholder participation and consultation–emphasized in many policies–still aren't being translated into concrete actions. In fact, many large-scale projects that are likely to have a great impact on a huge number of people are still being designed with little public scrutiny-or even in secret. *Thailand's planned water-grid project, which could involve water transfers on a massive scale, is a prime example.* It also has to be recognized that governments often fail to assess the benefits of proposed projects in relation to their costs, and that they certainly don't present such cost-benefit analyses to the public as information or for discussion.[956]

Water planning in the grid is largely expert-driven with preconceived procedures and results. There is scant regard for decision-making that is based on negotiations and enhanced relations with stakeholders and community groups, yet project planning continued through 2005, when the Master Plan on the grid was completed without any consideration of a smaller scale and less environmentally intrusive project. The Master Plan ostensibly included 13 grids with priority given to the Nam Ngum tributary tunnel, diverting water

953 *Id.*

954 International Water Management Institute, "Water Governance in the Mekong Region: The Need for More Informed Policy-Making (Water Policy Briefing Issue No. 22, November 2006) 2.

955 *Id.* at 3.

956 *Id.* at 1 (emphasis added).

from the Lao PDR under the Mekong to Northeast Thailand.[957] With the fall of the Thaksin government in 2006, the Thai water grid was presumed dead. However, in February 2008, the new prime minister, within a week of entering office, announced a plan to divert water from the Mekong River "through underground tunnels to Loei and Udon Thani, where reservoirs will act as distribution centers to send the water on to farms in other provinces during the dry season. The water will be transported through small pipelines."[958]

On 1 February 2010, reports indicated that Thai Natural Resources and Environment Minister Suwit Khunkitti planned to dust off the Thai water grid, a plan he initiated in 1995 when he was Agriculture and Cooperatives Minister, and allocated budget funds to study potential sites and gage stakeholder and community opinions about the plan.[959] The National Water Resources Committee planned to discuss the impacts of the grid's water diversions on all Thai river basins.[960]

In concert with the renewed interest in the Thai water grid and agricultural water redistribution, a feasibility study was conducted on the proposed Kong-Loei-Chi-Mun project by the Thai Royal Irrigation Department (rid) from 2008 to 2012.[961] In the study, the RID proposed that water would be withdrawn from the Mekong mainstream via tunnels under the Loei River, stored at the Huai Luang and Ubol Rattana reservoirs, then diverted via canals into the Chi and Mun rivers, where it would finally be used for irrigation, bolstering the existing flows in the Chi and Mun rivers.[962] The project gained further urgency in 2015, when severe drought plagued northeast Thailand, and prompted a more extensive feasibility study in 2016.[963] The study (being conducted by Panya

957 Hirsch and Jensen, *supra* note 180, at 150.

958 Ancha Charoenpo, *Megaproject is "a top priority" for new govt: Big plans for diversion of water from the Mekong* (4 February 2008) Bangkok Post.

959 Apinya Wipatayotin, *Water Grid Scheme Lashed* (1 February 2010) Bangkok Post.

960 *Id.*

961 Thu Suong, *Mekong Basin Stirs up Region: Thai Water Diversion Project Could have Mega Risks*, (8 July 2016) Mekong Eye <https://www.mekongeye.com/2016/07/08/mekong-basin-stirs-up-region-thai-water-diversion-project-could-have-mega-risks/> (accessed 20 June 2017).

962 *Id.* The name Kong-Loei-Chi-Mun is thus derived from the watercourses that the project utilizes. The Loei is a Mekong tributary that contributes to regional agriculture and is considered to have ample flows to do so. The Chi is another Mekong tributary, entering the mainstream downstream of the Loei, and is not endowed with a comparable ability to sustain extensive agricultural use (at least in the eyes of the RID). The Mun, in turn, is part of the Chi sub-basin.

963 *Id.*

Consultants on behalf of the RID) was scheduled for completion in December 2016, and seeks to ascertain the economic viability of the project, conduct environmental and social impact assessments, solicit public participation, and ultimately determine whether it should be carried out.[964]

The Kong-Loei-Chi-Mun project has received its due of criticism from multiple stakeholders, who are understandably troubled by the negative externalities that would presumably occur from such a massive hydroengineering endeavor. Principally, villagers located near the mount of the Loei are concerned that the project will bring no benefits to them and could indeed adversely impact them by flooding lowland residential areas from diverting too much water into the Loei at one time and potentially widen the mouth of the Loei, causing erosion and land loss.[965] One local village has even posted homemade signs that express their opposition to the Loei surface expansion, supporting these sentiments by noting that government officials have merely informed them that the project will be implemented and touted increased water for irrigation, rather than soliciting their input and providing more detailed information.[966] Additionally, experts and NGOs have pointed out that the project could eventually lead to saltwater intrusion and water shortages downstream in Vietnam and Cambodia.[967]

This last transboundary concern would presumably subject the Kong-Loei-Chi-Mun project to the MRC procedures, at least those regarding prior notification. Indeed, any project that diverts water from the Mekong mainstream during the dry season would be subject to the PNPCA. However, Panya Consultants has countered that, based on the plan they are analyzing, Thailand's diversions from the Mekong will fluctuate according to the flow rate in the mainstream, with little to no water withdrawn during the dry season.[968] RID officials have also guaranteed that, if the feasibility study reveals sufficient transboundary impacts to warrant application of the Mekong Agreement, then the RID will follow MRC protocols.[969] The RID has also continued to maintain the benefits of the Kong-Loei-Chi-Mun project that will befall farmers in the

---

964 Mai Lan, *Diverting the Mekong River into Thailand: The Khong-Loei-Chi-Mun Project* (6 June 2016) Mekong Commons <http://www.mekongcommons.org/large-scale-irrigation-northeastern-thailand-reprised-khong-loei-chi-mun-project/> (accessed 20 June 2017).

965 *Id.*

966 Suong, *supra* note 961.

967 *Id.*

968 Lan, *supra* note 964.

969 *Id.* Predictably, prior to the conclusion of the study, the RID did not feel that diversions for the project will be significant enough to impact downstream users and did not expect the PNPCA to be necessary.

seasonally arid northeastern provinces, including an increase in the average annual income commensurate with farmers in the rest of the country, which they feel will naturally occur after impacts from seasonal and regional droughts are alleviated by the Kong-Loei-Chi-Mun project.[970] Of course, these arguments seem inconsistent: if the water is primarily needed by farmers during seasonal and extreme droughts, how would the Kong-Loei-Chi-Mun project provide relief if withdrawals are based off of flows in the Mekong, which may also be impacted by the same drought? Although there have been no updates on the status of the Kong-Loei-Chi-Mun project as of June 2017, the renewed interest in a Thai water grid-inspired scheme evidences that the "desert bloom syndrome is alive and well" in Thailand.[971]

4.3 The Legal and Scientific Imperatives for Demarginalizing Tributaries in the Mekong Legal Regime: Dams, Water Diversions, Biogeography, and the Evolving Contours of Climate Change

Perhaps the largest threats to the fragile LMRB ecosystem today are the unbridled construction of dams and proposed water diversions on the river's tributaries. Dams are planned for lower sections of the Mekong mainstream along with numerous projects on tributaries.[972] According to Buxton and others, "The construction of dams [in the LMRB] is the activity most likely to lead to conflict over water."[973] In 2002, Bonheur observed that dam construction for hydropower and irrigation is "the most sensitive issue of all water related projects in the Mekong Basin."[974] The Lao PDR, Thailand, and Vietnam have erected several dams on tributaries, and hydroelectricity is now the biggest export from the Lao PDR.[975] As Pearce points out, "Most of these dams have

970 *Id.* Officials note that the average annual income for farmers is 131,000 baht in Thailand as a whole, compared to 68,000 baht in the northeastern provinces, rendering a large portion of the population stuck in generational poverty.

971 Molle and Floch (2008b), *supra* note 925, at 12.

972 Michael Buxton et al., "Environmental Conflicts in the Mekong River Basin: Prevention and Resolution" (Report to the Environment Programme, Mekong River Commission, School of Social Science and Planning, RMIT University, Melbourne, Australia 2003) 65.

973 Buxton et al. ("Conflict Resolution"), *supra* note 340, at 30.

974 Neou Bonheur, National Report on the Prevention and Resolution of Environmental Conflicts in the Lower Mekong River Basin (Cambodia) 2 (2002).

975 *See generally,* Grainne Ryder, *Ending Vietnam's Hydro Threat to Cambodia's Mekong Tributaries: Why Power Sector Reform Matters* (Probe International, November 2002); IRN (2004), *supra* note 888, at 10.

proved problematic in one way or another."[976] In addition to Yali Falls dam, Pak Mun dam in Thailand and Nam Theun dam in the Lao PDR have already severely impacted indigenous peoples and local fisheries. In the greater Sesan/3S basin, there are now seven commissioned hydroelectric dams in the Vietnamese Sesan cascade, and seven more planned in the Cambodian portion of the basin (including Lower Sesan 2, which is under construction)."[977]

Construction began on the Sesan 3 project, located some 15 km downstream of Yali Falls, in May 2001 and was completed in 2006.[978] Construction on this dam started before an EIA had been agreed to or conducted. The Sesan 3 project features a 70-m high concrete gravity dam and two 96-m penstocks to a surface power station with two units.[979] The dam has a capacity of 260 MW and creates a 3.4 km^2 reservoir with storage of 92 million m^3.[980] Mean annual flow is 274 m^3/sec and the turbine discharge is 486 m^3/sec.[981]

In 2003, EVN began construction of the Pleikrong hydropower project located on the Krong Poko tributary, about 3 km upstream of the confluence with the Dak Bla River, approximately 20 km west of Kontum.[982] Pleikrong is the only dam upstream of Yali Falls. The Pleikrong dam became operational in 2009. It features a 71-m roller-compacted concrete dam and a 100-m long embedded penstock to a surface power station with two units and an installed capacity of 100 MW.[983] Mean annual flow is 128 m^3/sec and the turbine discharge is 367.6 m^3/sec. The dam's reservoir covers an area of 53 km^2 and has a storage of 950 million m^3, corresponding to some 24 percent of the average runoff.[984] Regulation of the Krong Poko River will increase the average annual power generation by some 290 gigawatt hours for downstream projects.[985]

The Sesan 3A hydropower project, located 10 km downstream of the Sesan 3 dam, features a 34-m high concrete gravity dam and a surface power station

976 Pearce, *supra* note 373, at 102.

977 IRN, *Damming the Sesan River: Impacts on Cambodia and Vietnam* (October 2002) <https://www.internationalrivers.org/resources/damming-the-sesan-river-impacts-in-cambodia-and-vietnam-2673> (accessed 19 September 2017); Rieu-Clarke and Gooch, *supra* note 423, at 201.

978 SWECO Groner, *supra* note 803, at 42.

979 *Id.*

980 *Id.*

981 *Id.*

982 *Id.* at 41.

983 *Id.*

984 *Id.*

985 *Id.*

with two units at the foot of the dam.[986] Installed power capacity of the Sesan 3 project is 96 MW and its 8.5 km^2 reservoir creates an active storage of 4 million m^3.[987] Construction of the dam started in April 2003 and it was in operation in by late 2006.[988] Mean annual flow is 284 m^3/sec and the turbine discharge is 500 m^3/sec.[989]

Finally, excavation for the Sesan 4 dam, a 74-m high rock fill dam located some 22 km downstream from the Sesan 3A hydropower project, began in 2004.[990] The Sesan 4 dam became operational in 2010. The project features an embedded penstock adjacent to a surface power station with three units.[991] The reservoir creates an area of 58 km^2 with storage of 264 million m^3, corresponding to 3 percent of the annual flow.[992] Mean annual flow is 328.9 m^3/sec, and the turbine discharge is 719 m^3/sec.

To offset the impacts of hydropower development in Vietnam on the Cambodia part of the Sesan, EVN began construction of the Sesan 4A re-regulation reservoir in 2004, approximately 5 km downstream of the Sesan 4 power station and 1 km from the Cambodian border.[993] Located at the terminus of Vietnam's Sesan cascade, this 1.73 km^2 reservoir is designed to provide re-regulation of the intermittent outflow from the upstream hydropower dams, and consequently, to eliminate daily flow variations in Cambodia.[994] There is no installed generation capacity in the dam, which was completed in 2009.[995]

Elements of the Thai water grid were resurrected by the new Thai government in 2008 in the form of a RID project known as Kong-Loei-Chi-Mun.[996] In addition to the Kong-Loei-Chi-Mun project, another megaproject known as the Thai-Lao Water Friendship is still in the works, although little action has been taken in further assessing its feasibility since the mid-2000s.[997]

---

986 *Id.* at 42.
987 *Id.*
988 *Id.*
989 *Id.*
990 *Id.* at 43.
991 *Id.*
992 *Id.*
993 *Id.* at 46.
994 *Id.*
995 *Id* at 7.
996 Suong, *supra* note 961. *See* discussion under Section 4.2, *supra.*
997 *See* Aaron T. Wolf, Tiodd Jarvis, and Rejani Kunjappan, "Effective Negotiation for Transboundary Waters: A Skills Building Course", prepared for the MRC by authors, (April 2014) 126. *available at* <http://www.transboundarywaters.orst.edu/training/Workshop%20Materials/Mekong%20-%202014-%20Effective%20Negotiation%20TransWaters%20-%20Skills%20Building%20Course.pdf> (accessed 19 September 2017).

If the Thai-Lao Water Friendship is honored, tributaries in the Lao PDR will be subject to large water diversions which will have significant effects on the sub-basins, including the Nam Ngum, Se Bang Fai, and Se Bang Hiang tributaries. Although the water grid proper seemingly faded with the end of the Thaksin era, the scheme of transferring water from Lao PDR tributaries, siphoning it under the Mekong via tunnels and irrigating Northeast Thailand reappeared in 2007.[998] The Lao PDR tributaries water transfer plan includes part of the water grid first tested in 1998 which envisaged building two dams on the Se Bang Hiang near its Mekong confluence from which 3.3 billion m^3 of water would be diverted.[999] As long as elements of the Thai water grid are part of the Thai and Lao PDR governments' water resources stratagem, these tributaries and their sub-basins remain threatened.

The Yali Falls Dam, developed and operated in ways that violate customary international law and the 1995 Mekong Agreement, affected the Sesan all the way downstream to its confluence with the Mekong in Stung Treng Province. Arguably, the 1995 Mekong Agreement was not in force at the onset of construction of Yali Falls dam in 1993, but this did not release Vietnam from its obligations under customary international law, which is meant to prevent such abuses. Further, under Article 10 "Emergency Situations," the 1995 Mekong Agreement applied to any subsequent Vietnamese operations at Yali Falls after effectuation of the treaty.

As seen earlier, the three main rules of customary international law concerning shared freshwater, equitable and reasonable utilization, prevention of significant harm, and prior notification, are codified in the UN Watercourses Convention which entered into force on 17 August 2014. For instance, equitable and reasonable use acknowledges that upstream and downstream states each have a right to share a river's resources. This principle is a customary norm embodied in Article 5 of both the UN Watercourses Convention and the 1995 Mekong Agreement. Article 5 calls on the LMRB states "[t]o utilize the waters of the Mekong River system in a reasonable and equitable manner in their respective territories, pursuant to all relevant factors and circumstances."

International law requires the LMRB states to consider the effects of their use of a watercourse on downstream riparians and not to cause significant harm.[1000] Article 7 of the UN Watercourses Convention, "Obligation Not to Cause Significant Harm," requires that:

998 SWECO Groner, *supra* note 803, at 7.

999 Rivers Watch, "Laos-Thai Friendship Water Diversion Project" (2005).

1000 UN Watercourses Convention, *supra* note 18, art. 7; 1995 Mekong Agreement, *supra* note 1, art. 7.

1. Watercourse States shall, in utilizing an international watercourse in their territories, take all appropriate measures to prevent the causing of significant harm to other watercourse States.
2. Where significant harm nevertheless is caused to another watercourse State, the States whose use causes such harm shall, in the absence of agreement to such use, take all appropriate measures, having due regard for the provisions of articles 5 and 6, in consultation with the affected State, to eliminate or mitigate such harm and, where appropriate, to discuss the question of compensation.

In the *Gabcikovo-Nagymaros* case, the ICJ confirmed that states are not to injure the environment outside their borders.[1001] The purpose of the Mekong legal regime is "to minimize the harmful effects that might result from natural consequences and man-made activities."[1002] Article 7 of the 1995 Mekong Agreement obligates member states to "make every effort to avoid, minimize and mitigate harmful effects that might occur to the environment, especially the water quantity and water quality."[1003] Further, if a state is notified—as Vietnam was in 2000[1004]—that it is causing "substantial damage," a term undefined in the 1995 Mekong Agreement, it is obligated to "cease immediately the alleged cause of harm."[1005] Vietnam has neither ceased its dam operations nor formally assessed the harm[1006] that the Yali Falls dam caused to Cambodian villagers. The downstream effects of the development and operation of the Yali Falls dam exceed an equitable and reasonable use of the Sesan River and violate the obligation to prevent significant harm.

International watercourse law requires that timely notice (six months) accompanied by adequate technical information be given in advance of planned activities that could impact downstream states.[1007] One of the key components of notification is an EIA. Here again, the 1995 Mekong Agreement is deficient, requiring only that "[o]n tributaries ... intra-basin and inter-basin diversions shall be subject to notification to the [MRC] Joint Committee."[1008] EIAs are not legally required in the 1995 Mekong Agreement for mainstream projects, let alone for intrabasin water use or interbasin water diversions on tributaries.

1001 *Gabcikovo-Nagymaros* case, *supra* note 474, para. 53.

1002 1995 Mekong Agreement, *supra* note 1, art. 1.

1003 *Id.*, art. 7.

1004 MRC, Fax from the Mekong River Commission and accompanying Report on visit to Ialy [Yali] Hydropower Project 20–21 April 2000 (15 May 2000).

1005 1995 Mekong Agreement, *supra* note 1, art. 7.

1006 *Id.*, art. 8.

1007 UN Watercourses Convention, *supra* note 18, arts. 11–12.

1008 1995 Mekong Agreement, *supra* note 1, art. 5(A).

As noted in Section 2.2.5 of this book, "Notification" in the 1995 Mekong Agreement means "[t]imely providing information by a riparian to the [JC] on its proposed use of water according to the format, content and procedures set forth in the Rules for Water Utilization and Inter-Basin Diversions under Article 26."[1009] The definition of "notice" in the Procedures for Notification, Prior Consultation and Agreement, which were approved in 2003 after the completion of the Yali Falls dam, mirrors that in the 1995 Mekong Agreement. According to Osborne, "it is clear that no effort was made by Vietnamese authorities to notify their Cambodian counterparts of what was happening."[1010] After the initial onslaught of pulses from the dam, a notification system was established whereby Vietnam "faxes" notices of intent to release water to Cambodia. On several occasions, advance warning was not given, or was at best tokenistic, in that a fax was sent after the release had taken place.[1011] Gaps in the notice process continue to be a problem.[1012] The lack of notice of construction and operation of the Yali Falls dam was a contributing factor in the deaths, property losses, and other downstream impacts on the Sesan River in Cambodia.

"A fundamental rule of interpretation of legal texts is to evidence the original will of the drafters."[1013] The 1995 Mekong Agreement governs the "Mekong River Basin,"[1014] but the treaty leaves this term undefined. Drawing from the schism between the intent of the Agreement evidenced in Articles 1–3 and the unsustainable development so pervasive in the Yali Falls disaster and the proposed diversions in the Thai water grid, it is starkly apparent that a holistic approach to water resources management has not been applied in the LMRB. Although the MRC is responsible for management of the mainstream, tributaries are managed by each basin state independently through its NMC according to each country's national agenda. Because many tributaries flow through more than one country, it is inconsistent to treat international tributaries disparately from the mainstream. Simply stated, tributaries are insufficiently covered by the 1995 Mekong Agreement and its procedures.

Although the concept of a community of interest is not spelled out or directly mentioned in the 1995 Mekong Agreement, certainly a notion of the concept is implied. The term "community of Mekong nations" is noteworthy in paragraph six of the Preamble and calls for joint cooperation between the LMRB states. The *River Oder* and *Gabcikovo-Nagymaros* cases arguably strengthen

---

1009 *Id.*, Ch. II, para. 7.

1010 Osborne (2004), *supra* note 208, at 32–33.

1011 *Id.* at 33.

1012 Down River, *supra* note 16, at 54.

1013 Wohlwend, *supra*, note 513, at 2.

1014 1995 Mekong Agreement, *supra* note 1, arts. 1–3.

the concept of a community of interest for non-navigational uses. McCaffrey notes that it is not insignificant that in the *Gabcikovo-Nagymaros* case, there is recognition by the ICJ that all riparian states have interests in an international watercourse, which he argues constitutes an effective repudiation of absolute territorial sovereignty.[1015] McCaffrey argues, and rightly so, that the very existence of the 1997 UN Convention evidences a strengthening of the concept of a community of interest in a basin and ascribes to its adoption "a confirmation of the development of international law in the direction of requiring that riparian states recognize the rights of other riparians in shared freshwater resources."[1016] According to McCaffrey, in the *Gabcikovo-Nagymaros* case the ICJ:

> [A]pplies this [concept] to the case at hand ..., in which it finds that Czechoslovakia, by unilaterally damming the Danube (a shared resource) at a point at which it was wholly within Slovak territory, thereby deprive[ed] Hungary of its right to an equitable and reasonable share of the natural resources of the Danube.[1017]

He characterizes the ICJ's recognition of the concept of a shared watercourse as "a strong endorsement of equitable utilization as a norm of customary international law, [that] should remove any lingering doubt about the status of that principle."[1018] In this sense, *the River Oder* case and its progeny, the *Gabcikovo-Nagymaros* case, implicitly extend the jurisdiction of the MRC to the Mekong "basin's" tributaries, but for the legal effects of Article 5(A). In some small way, then, perhaps the concept of "the community of Mekong nations," is implicitly linked to or is a corollary of the concept of a community of interest, which is incorporated into the customary norm of equitable utilization in Articles 5 and 26 of the 1995 Mekong Agreement, irrespective of Article 5(A)'s marginalization of tributaries.

The marginalization of tributaries in the Mekong legal regime evidences an unwillingness of the LMRB states to follow the proper channels of international watercourse law as memorialized in the framework of the UN Watercourses Convention and presupposed in the sustainable development paradigm of the 1995 Mekong Agreement itself. Rather, the countries are pursuing hydrosovereignty. This state-centric approach fails to provide a satisfactory framework for

1015 McCaffrey (2001), *supra* note 520, at 260.

1016 *Id.*

1017 *Id.* (Internal quotations omitted).

1018 *Id.*

dealing with events occurring at a regional scale. Indeed, the very establishment of the MRC seemingly clashes with the sovereignty values of the LMRB states.

According to Carrard, "[c]ompared with the relatively [underdeveloped] mainstream, Mekong tributaries are commonly subject to development and disruption of their natural [instream and environmental] flow regimes."[1019] Although the 1995 Mekong Agreement distinguishes between the mainstream and its tributaries for management purposes, the treaty lacks effective legal mechanisms for regulating development on tributaries.[1020] Consequently, tributaries are marginalized within the 1995 Mekong Agreement, and the legal regime allows the LMRB member states to circumvent requirements for cooperative transboundary governance of the LMRB.[1021]

Although the notification process applies to tributaries, this procedure does not legally require any mandatory review, deliberation, or prior consultation relative to planned measures by the LMRB states. Consequently, megaprojects, such as hydropower dams and water diversions on tributaries, are not subject to any joint cooperative development agreements. Mainstream interbasin diversions in the dry season are subject to agreement by all LMRB states through the JC. This dichotomy provides insight into why larger projects are undertaken on tributaries rather than the Mekong proper. Ironically, tributary dams and water diversions clearly impact the water quantity and quality of the entire LMRB. The disparate treatment of tributaries in the Mekong legal regime is at best questionable in the management of the river's hydrological regime because diversions on tributaries in the dry season could, in effect, have the same impact as diversions from the mainstream on total water flows, albeit maybe on a smaller scale.[1022] Since mainstream flows are dependent on the variable flows from tributaries, consideration should be given to including tributaries in the Mekong legal regime to ensure maintenance of minimum flows on the mainstream as required in Article 6 of the 1995 Mekong Agreement. Vietnam has effectively done this by joining the UN Watercourses Convention, and so should Cambodia, Lao PDR, and Thailand. Otherwise, the limited extent to which planned measures on tributaries are subject to basin-wide management marginalizes the Mekong's arteries and renders the 1995 Mekong Agreement hydrologically artificial.

1019 Carrard, *supra* note 10.

1020 *Id.*

1021 *Id.*

1022 Browder, *supra* note 250, at 255.

Drawing on the conflicts over the construction and operation of the Yali Falls dam and proposed water diversions of the Thai water grid, if the proposition is accepted that unregulated development on tributaries is a potential flashpoint for cross-border conflict and far-reaching harmful environmental and social effects in the LMRB, a significant deficiency in the 1995 Mekong Agreement is revealed. It treats transboundary tributaries as if they are confined to one jurisdiction. There are no legal mechanisms for regulating dam development and water diversions on tributaries and, in effect, they are not managed holistically as part of the wider LMRB. Without legal mechanisms to control development on tributaries, water resources will be over-utilized by hegemonic states such as Thailand and Vietnam and underutilized by the poorer states of Cambodia and the Lao PDR. This asymmetry is further skewed by Vietnam's upstream position and hydro-hegemony.[1023] The indigenous peoples of both Cambodia and the Lao PDR derive neither inter- or intra-generational equity, nor any long-term environmental, social, or economic benefits from dams such as Yali Falls or the proposed water diversions in the Thai water grid. This constitutes "unsustainable basin management,"[1024] which is contrary to the requirement of protection of a watercourse's ecosystem in Article 20 of the UN Watercourses Convention and Article 3 of the 1995 Mekong Agreement. Because of these practices, one commentator has dubbed the 1995 Mekong Agreement as the "toothless guardian" of the LMRB.[1025]

The Yali Falls dam and the Thai water grid case studies evidence that the national interests of economic development are advanced at the expense of basin-wide water governance and holistic water resources management in the LMRB. According to Hirsch and Jensen, "[i]n the Vietnamese context, with a rapidly growing economy and intense agricultural and industrial development, dam building is considered a fundamental component of national plans to boost energy production and increase wealth."[1026] Likewise, the water grid is lauded as a means to alleviate poverty, quench thirst in Bangkok and green Thailand's northeast region.[1027] The 1995 Mekong Agreement indeed encourages promotion of national interests by allowing the LMRB states "[t]o cooperate on the basis of *sovereign equality* and *territorial integrity* in the utilization

1023 *See generally,* Zeitoun and Warner, *supra* note 332.

1024 Rix, *supra* note 800, at 104.

1025 Barry Wain, "Mekong River: River at Risk—The Mekong's Toothless Guardian", *Far Eastern Economic Review* (26 August 2004).

1026 Hirsch and Jensen, *supra* note 180, at 146.

1027 *Id.* at 150.

and protection of the water resources of the Mekong River Basin."[1028] Moreover, the scope of the agreement allows:

> Parties … [to] enter into bi- or multi-lateral special agreements or arrangements for the implementation and management of any programs and projects to be undertaken within the framework of this Agreement, which agreements shall not be in conflict with this Agreement and shall not confer any rights or obligations upon the parties not signatories thereto, except as otherwise conferred under this Agreement.[1029]

The 1995 Mekong Agreement then, allows the LMRB states to enter into bi- or multilateral agreements *with each other and other states*. Read together, Articles 4, 5(A), and 38 of the 1995 Mekong Agreement seemingly provide ample avenues of creativity for justifying unilateralism or bilateralism in planned measures on tributaries, irrespective of the politics of scale. Consequently, it is no coincidence that most of the major international legal controversies in the LMRB relate to existing and planned measures on Mekong tributaries. This is because the decision making regarding tributaries is guided by the principle of hydrosovereignty, which in this case is essentially tantamount to the Harmon Doctrine.[1030] Read in context together, Articles 4, 5(A), and 38 provide enough flexibility to allow collateral agreements that can be used to conveniently sidestep, yet remain consistent with, the 1995 Mekong Agreement for large-scale projects on transboundary tributaries that ultimately affect water quantity and quality and instream flows in the Mekong mainstream.

Ironically, as previously noted in Chapter 3 of this book, Vietnam argued for a continuation of the unanimity rule in the negotiations of the 1995 Mekong Agreement,[1031] relying primarily on the doctrine of absolute territorial integrity which provides that "the upstream state may do nothing that might affect the natural flow of the water into the downstream state."[1032] This was a natural progression in the negotiations during the evolution of the tributary rule,[1033] as Vietnam is a downstream state on the Mekong mainstream but an upstream state on the Sesan River. As noted in Section 2.2.3 of this book, like the Harmon Doctrine, absolute territorial integrity has generally been rejected

1028 1995 Mekong Agreement, *supra* note 1, art. 4 (emphasis added).

1029 *Id.*, art. 38.

1030 *See generally*, McCaffrey, *supra* note 9, at 112–114.

1031 Browder, *supra* note 250, at 245–246.

1032 McCaffrey, *supra* note 9, at 126.

1033 1995 Mekong Agreement, *supra* note 1, art. 5(A).

as too extreme for modern state practice and inconsistent with equitable and reasonable utilization.[1034]

Although tributaries are explicitly referred to in Article 5(A) of the 1995 Mekong Agreement, they are only subject to rules of notification for planned measures or water diversions, irrespective of scale. In effect, tributaries are exempted from the more stringent requirements for mainstream planned measures in the 1995 Mekong Agreement, thereby taking these arteries outside the treaty or marginalizing them. Consequently, it is clear that within the Mekong legal regime they are not managed holistically.

A study by the River Basin Science and Policy Center at the University of Georgia (US) found that impoundments can severely affect large and small tributaries in five major ways.[1035] Not surprisingly, all five of these effects are observable in the LMRB. First, reservoirs and water diversions alter natural flow regimes.[1036] Mike Kensler of Auburn University's Water Center has observed that:

> More than any other factor, it is the hydrological component of naturally variable river flow that determines river health and the viability of the complex web of relationships that comprise river systems. Naturally flowing, unaltered rivers are inherently variable. This variability includes the magnitudes of low, high, and intermediate flows, as well as the seasonal timing, duration, frequency, and rate of change of flows. In unaltered rivers, and in altered rivers that are managed to mimic the natural flow regime, flows fluctuate dynamically within a predictable range and at predictable times throughout seasons, years, and decades.
>
> Ecologically intact river flow patterns provide a number of important functions and services. Naturally occurring flows enable sediment transport, nutrient uptake, the viability of the food web, and habitat maintenance and creation, thereby protecting biodiversity, water quality and water availability.
>
> Plants and animals have co-evolved with rivers over time to use flow variability to the greatest advantage. For example, flow patterns are critical to the maintenance of river-floodplain ecosystems. Native plants and animals along and within rivers have adapted to frequent overbank

1034 McCaffrey, *supra* note 9, at 133–135.

1035 Gail Cowie (ed.), "Reservoirs in Georgia: Meeting Water Supply Needs While Minimizing Impacts" (University of Georgia River Basin Science and Policy Center, May 2002) 4–13.

1036 *Id.* at 4.

> flooding during winter months, and low flow periods in late summer. Movements of fishes, availability of food, and reproductive success (courtship, spawning, hatching) are highly attuned to flooding, low flows, and variations in between, occurring at predictable times of the year.
>
> Frequent high flows clear streambeds of fine sediment and deposit this sediment on floodplains where it enriches forest soil and maintains forest productivity. Overflow into floodplains also removes excessive nutrients dissolved in river water as vegetation and soils in floodplains capture nitrogen and phosphorus during periods of flooding. High flows scour flood habitat and create new spaces for native, flood-adapted seedlings to become established.
>
> Floodplain forests are refuges for many economically valuable fish species during high flows. In these floodplain refuges, food is abundant, fishes are safe from the swift-flowing main channel, and growth rates are increased. During low stream flow periods, seeds of floodplain forest vegetation will germinate in the dry soil.
>
> The key learning is that natural variability of river flows, meaning the magnitude, timing, frequency, duration, and rate of change of flows in rivers and streams, is fundamentally important to maintaining healthy river-floodplain forest ecosystems and perpetuating the ecosystem services they provide.[1037]

The natural flow regime, then, is the "master variable" of a river that controls many ecosystem traits, including biogeography, biodiversity, and species distribution.[1038] Seasonal and annual flow variations are needed in tributaries to provide energy, habitat, and biodiversity. Flooding connects tributaries to wetlands and floodplains, providing feeding grounds and refugia from prey for aquatic biota and facilitating exchange of organic materials. In essence, the lack of hydrologic variability on a tributary hydrograph, signaling minimum flow or flat-line instream flow, may foretell the death knell for that river, similar to the flatline on an electrocardiogram or electroencephalogram. Cowie describes the effects of altered flow regimes on tributaries:

> When natural flow regimes are altered, the basic function of riverine ecosystems is diminished. The environmental impact of reservoirs on

1037 See email from Mike Kensler, Director of Outreach Programs, Auburn University Water Center, to author (11 May 2010, 9:24 a.m. CST) (on file with author).

1038 N.L. Poff et al., "The Natural Flow Regime" (1997) 47 *Bioscience* 769–784; Cowie, *supra* note 1035, at 4.

> streams and rivers is dramatic because reservoirs are the only structures that can simultaneously and thoroughly alter water level fluctuations, continuity of flow, and the amount of water in the system.
>
> ....
>
> By its nature, a river is a dynamic ecosystem, with water levels rising and falling with the rains .... This pattern changes from year to year and in different parts of the state, and each river has its own pattern. While effects of reservoirs on flow patterns depend on the purpose for which the reservoir was constructed, differences in flow patterns after dam construction are often dramatic.
>
> While the effects of reservoirs are most dramatic in the case of large dams on large rivers, small dams on small streams have significant local effects that, in aggregate, are very large.[1039]

Second, impoundments fragment the tributary system and cause loss of continuity.[1040] Impoundments can severely damage tributary health and functions by fragmenting habitat. Multiple impoundments along a tributary, such as those on the Sesan, can cause even more fragmentation, altering connectivity and hindering important natural functions such as fish migration and spawning. Cowie explains the concept of loss of continuity:

> The very nature of free-flowing streams and rivers is determined by a continuous connection of water and the associated energy, sediments, nutrients, and biota, from headwaters to the sea. The food web of rivers, for example, is based on inputs from tributary streams, and downstream areas are sustained by the flow of water and transported materials. Dams placed across channels, whether large or small, block these connections, isolating upstream segments from the rest of the system.
>
> Multiple reservoirs can result in highly fragmented systems.[1041]

While limited attention has been given to mainstream flows in the Mekong regime procedures, the impact of fragmentation on tributaries has not been a management concern.

1039 Cowie, *supra* note 1035, at 4–5.

1040 *Id.* at 4 and 6.

1041 *Id.* at 6; *See generally,* Nikolai Sindorf and Trang Dangthuy, "Aquatic Ecosystem Connectivity, Ecosystem Processes and Watershed Integrity under Mekong Dam Development" (MRC International Conference on Watershed Management, Chiang Mai, Thailand, 9–11 March 2011). (analyzing the concept of aquatic ecosystem connectivity on the Mekong and its tributaries).

Third, impoundments on tributaries interrupt the downstream transport of sediment and alter stream channels.[1042] In their natural state, Mekong tributaries form channels that balance sediment supply and runoff. These predevelopment channels are stable. Dams and water diversions on tributaries often produce hydrologic changes such as stream widening and bank erosion, stream downcutting, loss of riparian tree canopy, increased erosion, and reduced sedimentation. When these developed channels stabilize, they are entrenched. Cowie aptly notes the imbalance caused by reservoirs:

> Reservoirs interrupt the downstream transport of sediment and floodwaters, resulting in a stream channel that is no longer in balance with its sediment and flow regime. The effects of this imbalance change with increasing distance from the dam. In the stream near the dam, channel degradation through scouring of sediments is most common. Because reservoirs trap sediment, water flowing from the reservoir carries a reduced sediment load. During moderate to high flows from the reservoir, the "sediment-starved" water leaving the reservoir erodes the channel bed and banks until a balanced sediment load is achieved, with several consequences. One is bed coarsening, a relative shift of the sediment particle size distribution from finer materials to larger materials. The quality of habitat for fishes and invertebrates declines where the bed becomes "armored" or hardened by loss of moveable sediments. Another consequence is incised, unstable stream channels and accelerated rates of bank erosion below reservoirs, which can cause loss of property.
>
> Sedimentation and increased elevation of the stream channel, known as channel aggradation, most commonly occurs farther downstream. Sediment delivered from degrading upstream areas and by tributary streams accumulates in the channel and cannot be naturally removed because the reservoir restricts the frequency and size of flood peaks. This accumulation of sediment occurs in addition to sediment inputs to the system during reservoir-related land clearing and dam construction. Sediment accumulation degrades habitat by filling pools and clogging riffles and reduces a stream's capacity to transport high flows, setting the stage for more severe flooding than would otherwise have occurred.[1043]

Fourth, impoundments on tributaries alter the physical and chemical processes which disrupt rivers' "self-purifying systems,"[1044] thereby diminishing

1042 Cowie, *supra* note 1035, at 7.

1043 *Id.* at 7–8.

1044 *Id.* at 8.

their ability to assimilate waste and maintain water quality. If water released from dams is from the deep part of a large reservoir such as Yali Falls, deadly blue-green algae can be released, potentially causing acute downstream water quality impairment. Recent research demonstrates that small tributaries provide a first line of defense against nonpoint source pollution through nutrient removal, much of which occurs in headwaters.[1045]

Fifth, cumulative impacts of multiple reservoirs, such as those in the Sesan cascade pose significant concerns for the function of that tributary's ecosystems. In the US, cumulative impacts are defined in federal regulations as "the impact on the environment which results from the incremental impact of the action when added to other past, present, and reasonably foreseeable future actions regardless of what agency ... or person undertakes such other actions."[1046] The cumulative impacts of multiple reservoirs on Mekong tributaries have yet to be examined in detailed scientific investigations, much less understood. However, it is clear that "assimilative capacity ... is governed to a large extent by flow and can be compromised by upstream reservoirs.[1047] "We are just beginning to recognize and assess the cumulative impacts of the thousands of dams that exist."[1048]

Mekong tributaries should also be considered in light of their biogeography, the analysis of spatial distributions of organisms.[1049] With an annual harvest of twice the catch in the North Sea, the Mekong and its tributaries yield more fish than any other river basin in the world.[1050] Ranking behind only the Amazon and Congo basins, the LMRB is home to more than 1,500 different fish species and is one of the world's most biodiverse basins.[1051] Tributaries play an important role in biodiversity by contributing to the longitudinal organization of biota in river systems.

1045 R.B. Alexander et al., "Effect of Stream Channel Size on the Delivery of Nitrogen to the Gulf of Mexico" (2000) 403 *Nature* 758, 758–761; "Lost Linkages and Lotic Ecology: Rediscovering Small Streams" in M.C. Press et al. (eds.), *Ecology: Achievement and Challenge: 41st Symposium of the British Ecological Society* (jointly sponsored by the Ecological Society of America, held at Orlando, Florida, USA, 10–13 April 2000; Cambridge University Press, 2001); B.J. Peterson et al., "Stream Processes Alter the Amount and Form of Nitrogen Exported from Small Watersheds" (2001) 292 *Science* 86–90.

1046 40 CFR 1508.7, *supra* note 11.

1047 Cowie, *supra* note 1035, at 13.

1048 *Id.*

1049 Bates and Jackson, *supra* note 22, at 66.

1050 Josefin Gooch, *The Mekong Agreement—In Light of International Principles Relating to Sustainable Use of Transboundary Watercourses* (Masters Thesis, Faculty of Law, University of Lund, Sweden, 2005) 9.

1051 *Id.*

River systems, such as the Mekong, can be thought of as a series of confluences or tributaries (nodes), separated by links.[1052] At each tributary's confluence with the main river channel, water and sediment enter the mainstream promoting hydrological, hydraulic, geomorphic, and ecological adjustments.[1053] According to Hoey et al., "[t]hese contributions are especially significant where the mainstream is regulated [through dams or diversions] such that understanding tributary impacts is especially important for the management of regulated rivers."[1054] Water quantity and quality in the mainstream is largely determined by volumetric contributions from tributaries. Consequently, degradation in tributaries is often reflected in the mainstream. However, only recently have tributaries and their attendant confluences been recognized as important in organizing the ecology of large rivers by creating complex habitats favored by fishes.[1055] On watercourses with extensive dams and diversions, "tributary processes dominate hydrologic and geomorphic conditions."[1056] Anthropogenic impacts challenge tributaries and can eliminate lateral habitats and lead to a homogenization of aquatic communities.[1057] Research indicates that "dams homogenize flows mainly by reducing peak flows and increasing minimum flows, creating more uniform flow conditions."[1058] Homogeneous rivers promote homogeneous faunas.[1059] Tributaries enhance the local heterogeneity at the confluence with the main river channel.[1060] Brown and Coon reported "higher fish density and different community composition in tributaries when compared to the channel."[1061] According to Dinakaran and Anbalagan, "the potential importance of tributaries for mainstream communities is virtually

1052 T. Hoey et al., "The Importance of Tributary Sediment Supply for the Sediment Dynamics of a Large Regulated Gravel-Bed River" (abstract) 1 (2004) (on file with author).

1053 *Id.*

1054 *Id.*

1055 Lee Benda et al., "Confluence Effects in Rivers: Interactions of Basin Scale, Network Geometry, and Disturbance Regimes" (2004) 40 *Water Resources Research* WO5402, doi: 10.1029/2003WR002583.

1056 Peter B. Moyle and Jeffrey F. Mount, "Homogenous Rivers, Homogenous Faunas" (April 3, 2007) 104 *Proceedings of the National Academies of Sciences of the U.S.A.* (PNAS) 5711, 5712.

1057 S. Dinakaran and S. Anbalagan, "Anthropogenic impacts on aquatic insects in six streams of south Western Ghats" (2007) 7 *Journal of Insect Science* 1, 7.

1058 Moyle and Mount, *supra* note 1056, at 5711.

1059 *Id.*

1060 Lee Benda et al., "The Network Dynamics Hypothesis: How Channel Networks Structure Riverine Habitats" (2004) 54 *Bioscience* 413.

1061 D.J. Brown and T.G. Coon, "Abundance and Assemblage Structure of Fish Larvae in the Lower Missouri River and its Tributaries" (1994) 123 *Transactions of the American Fisheries Society* 718.

unexplored."[1062] "Tributaries may serve as important refugia for recolonising the main channel after disturbances (floods, droughts, pollution) and they are important habitats for early life stages of fish and invertebrates."[1063]

Particularly noteworthy is the research on tributaries, sediment sources and the longitudinal organization of biota along river systems by Rice, et al., which indicates that:

> Tributary confluences are sites along a main channel where, because of the introduction of water and (or) sediment, the water volume, bed sediment character and water quality of the mainstream can change abruptly .... Variables of importance to stream biota are affected and, in turn, it is suggested that the arrangement of tributaries and related features is an important control on the longitudinal organization of macroinvertebrate benthos at moderate spatial scales.[1064]

According to Steve Rice, one of the world's foremost authorities on tributaries and river confluences, there are two main arguments to be made in support of inclusion of tributaries in any legal regime such as the Mekong:

> 1. The distribution and health of in-stream biota in part reflects responses to abiotic environmental conditions. In this context, tributary confluences are important because they are places where recruitment of water, sediment, and organic matter can have a substantial impact on habitat in the recipient channel. Note that below dams, where mainstem disruption of water and sediment regimes means that tributary supplies of water and sediment are of enhanced importance, significant changes in aquatic communities are widely documented. In addition to these local impacts, tributaries are important for the large-scale longitudinal organisation of stream biota. They configure an abiotic framework within which confluence "nodes" and the intervening channel links are basic spatial elements of biotic organization in river systems.
> 2. In addition, tributaries and confluence zones (those parts of the main stem channel directly affected by tributaries) are important because

---

1062 Dinakaran and Anbalagan, *supra* note 1057, at 7 (internal citations omitted).

1063 *Id.*

1064 S.P. Rice, et al., "Tributaries, Sediment Sources, and the Longitudinal Organization of Macroinvertebrate Fauna Along River Systems" (April. 2001) 58 *Canadian J. of Fisheries and Aquatic Sciences* 824.

they provide unique habitats and support important ecological functions.

* Confluences can be associated with increased productivity due to tributary supply of supplemental nutrients, drift and detritus.
* Confluences are sites where the juxtaposition of distinctive upstream, downstream and tributary environments may be usefully exploited by mobile species.
* The juxtaposition of contrasting environments may also enhance particular ecological processes such as refugia use.
* The morphology and hydraulics of confluences are unusual and may provide unique ecological opportunities for some organisms. An example of relevance for … Mekong work might be the apparent use of confluences by threatened species of freshwater dolphin; e.g. the Irrawaddy river dolphin.

….

* Geomorphological and sedimentological adjustment of the main stem channel to tributary water and sediment fluxes can increase physical heterogeneity at confluences ensuring a wider array of environmental conditions per unit channel length than would otherwise be present. That is, habitat variability is likely to be greater along a stretch of river including a confluence than along a similar reach without a confluence. In general, biotic diversity increases with habitat diversity.[1065]

Rice's findings underscore the importance of Article 20 in the UN Watercourses Convention which requires protection of a watercourse's *entire* ecosystem including its tributaries.

Investigations in the Amazon demonstrated that "tributaries have a large, but local, positive effect on mainstem species richness at confluences."[1066] Biogeographical analyses in that large river basin "suggest that local mainstem electric fish diversity is enhanced by tributaries."[1067] Gascon and Smith cogently summarized the results of the Amazon tributary investigation:

1065 Email communiqué from Stephen Rice, Department of Geography, Loughborough University, England, UK, to author (19 May 2009, 2:18 a.m. PDT) (on file with author). *See generally*, S. Rice et al. (eds.), *River Confluences, Tributaries and the Fluvial Network* (Wiley, 2008).

(This volume is considered by many commentators as the premier treatise on tributaries and river confluences.).

1066 Claude Gascon and Michael Leonard Smith, "Where Rivers Meet" (2004) 305 *Science* 1922.

1067 Cristina Cox Fernandes et al., "Amazonian Ecology: Tributaries Enhance the Diversity of Electric Fishes" (2004) 305 *Science* 1960.

> The authors investigated the local effects of tributaries by comparing species diversity in ... three types of station[s] ... at each confluence (above, in, and below the tributary). Diversity was highest in the tributaries themselves, suggesting that the Amazon encounters distinct biogeographical units as it cuts across the basin. Fernandes and colleagues found that stations just below the tributary confluence points had significantly higher diversity than stations just above them.[1068]

This evidences, then, that major tributaries conjoining with the Amazon increase species diversity immediately downstream of the confluence, but not upstream. This "tributary effect" may be the result of: (1) nutrient and organic material from terrestrial sources and floodplain lakes carried into the mainstream that provides food and habitat resources for fish; and/or (2) differences in flow volume and water levels between the Amazon and its tributaries caused by basin-wide seasonal rainfall and flood stages variations.[1069]

What is becoming increasingly clear is that rivers are not homogeneous entities. Studies similar to those summarized above should be conducted on Mekong tributaries to determine if the "tributary effect" can be observed in the basin. Indeed, the character of the Mekong is shaped by the flow, quality, and type of tributaries that enter it. Tributaries are where the Mekong River is born. As Mary Crane, a former contractor for the US Agency for International Development in Kabul, so aptly notes, "In Dari, there is an expression: *qatra, qatra, darya mesha,* or 'drop by drop it becomes a river'".[1070] But perhaps T.S. Eliot said it most eloquently:

> The river itself has no beginning or end. In its beginning, it is not yet the river; in its end, it is no longer the river. What we call the headwaters is only a selection from among the innumerable sources which flow together to compose it. At what point in its course does the Mississippi become what the Mississippi means?[1071]

1068 Gascon and Smith, *supra* note 1066, at 1923.

1069 R. Lowe-McConnell, *Ecological Studies in Tropical Fish Communities* (Cambridge University Press, 1987) 382.; R.H. Meade et al., "Backwater Effects in the Amazon River Basin in Brazil" (1991) 18 *Environmental Geology and Water Science* 105; Fernandes et al., *supra* note 1067, at 1962.

1070 Mary Crane, "It Becomes a River: Afghanistan at a Crossroads" (2009) 26(1) *World Policy Journal* 95.

1071 T.S. Eliot, "Introduction," *reprinted in* Mark Twain, *The Adventures of Huckleberry Finn* (Norton Critical Edition, 1950).

The volumetric contribution from Mekong tributaries should also be considered through the evolving contours of climate change law and policy. The shifting patterns of climate change will likely put additional stress on water supplies in the iconic Hindu Kush-Himalaya region of Asia. No place is this more apparent than in the Tibetan Plateau where melting glaciers that source the Mekong mainstream are signaling the coming ecological crisis.[1072] Thorson points out that "[g]lacial retreat portends significant global justice consequences."[1073] She provides a prescient outlook:

> Initially, glacial retreat is likely to flood riverbeds with water as more water than the glaciers otherwise supply is released; following that, however, the absence of the glaciers portends dry river beds except as rain or snowfall may supply water. This means less freshwater storage, more potentially catastrophic floods from glacial lakes, and economic losses from industries reliant on a constant supply of water.[1074]

Zhao and Schell decry the attendant effects of the altered flow regimes:

> With glaciers melting away *faster than anyone predicted*, the people of China, South Asia, and Southeast Asia are confronting the prospect of diminished water resources. For, while irregular river flows may be accelerated in the near term by the melting ice, the long-term flows would be diminished.
>
> ....
>
> Simply put, the Tibetan Plateau's environmental crisis is a 2-billion-person problem .... As half of the Asian population rely on the mighty river systems for water–for drinking, irrigation for agriculture, industry, and hydropower–the dramatic shrinking of glaciers and permafrost will cause a degradation of the plateau's ecosystem and disrupt reliable sources of downriver water supply for much of the continent.[1075]

1072 *See generally,* Michael Zhao and Orville Schell, "Tibet: Plateau in Peril" (2000) XXV(3) *World Policy Journal* 171.

1073 Erica J. Thorson, "Sharing Himalayan Glacial Meltwater: The Role of Territorial Sovereignty" (2009) 19 *Duke J. Comp. and Intl. L.* 487.

1074 *Id.* at 491–492. (internal citations omitted).

1075 Zhao and Schell *supra* note 1072, at 171–172 (emphasis added).

Yushu Prefecture in Qinghai Province, the source region of the Mekong, contributes 15 percent of the water to the mainstream.[1076] If current trends continue, the Tibetan Plateau, will lose two-thirds of its glacial ice cover by 2100.[1077] The 50 percent average annual flow contribution of Mekong tributaries to the mainstream will become increasingly more important to the LMRB states as water quantity problems are exacerbated by this " looming environmental meltdown."[1078]

The *Lake Lanoux Arbitration* (*France v. Spain; Lake Lanoux* arbitration)[1079] reflects a "community of interest" approach on waterbodies, such as tributaries and glaciers, interconnected to international river systems. Thorson summarizes this arbitral decision's importance to the concept of interconnectivity of waterbodies such as tributaries, glaciers and meltwater:

> In addressing these issues, the arbitral tribunal implicitly spoke to the extent of international water law. The panel recognized, by way of background, prior to arriving at any substantive conclusions, that the waters of Lac Lanoux, though wholly situated within the sovereign territory of France, are nonetheless an element of an international water system because of the interconnectedness of the lake and at least one transboundary river system. The panel pointed out that "there is a rule prohibiting the upper riparian State from altering the waters of a river in circumstances calculated to do serious injury to the lower riparian State". By so stating, the Panel confirmed that the customary international water law regarding nonnavigational uses includes within its scope lakes *and other waterbodies* [such as tributaries, glaciers and meltwater] that, though wholly situated within a sovereign territory, feed an international watercourse.[1080]

While it is probably an exaggeration to equate the *Lake Lanoux* arbitration with a "water-system approach" for tributaries, glaciers, and meltwater, it did set the

1076 *Id.* at 174.

1077 *Id.*

1078 *Id.* at 171.

1079 12 UNRIAA 281, 62 RGDIP 79 (1958), both in the original French. For English Translations, see 24 ILR 101 (1961), 53 AJIL 156 (1959), and 1974 YBILC, vol. 2, Part 2, p. 194 (1976). *See generally,* Sands, *supra* note 616, at 463–464; McCaffrey, *supra* note 9, at 130 and 225–227.

1080 Thorson, *supra* note 1073, at 505 (internal citations omitted). (emphasis added).

stage for the 1966 Helsinki Rules which adopted the basin rubric.[1081] Thorson posits that the 1995 Mekong Agreement is not expansive enough to cover Tibetan glaciers, primarily because of the figurative "lopped headwaters"[1082] resulting from China's refusal to ratify the treaty.[1083] Like tributaries, glaciers and their attendant meltwater need to be factored into the Mekong legal equation.

4.4 Chapter 4 Conclusion

The Yali Falls dam and the proposed water diversions in the Thai water grid are controversial and dramatic examples of the transformative power of Article 5(A) in the 1995 Mekong Agreement, a rule that allows LMRB states to use water in Mekong tributaries basically as they please. Both of these case studies pointedly highlight the effects of unbridled and unregulated development within the LMRB, yet they are only two among dozens of megaprojects that divert water and alter flow regimes on Mekong tributaries. The MRC, and indeed, the 1995 Mekong Agreement itself, are becoming increasingly sidelined by Article 5(A) proving that the emerging challenges in holistic water resources management in the LMRB require not only political will, but also more importantly, a functional legal regime. The cases studies demonstrate that tension between the LMRB states' national interests is the most hydropolitically sensitive obstacle to basin-wide governance. Otherwise, if this is not the case, Article 5(A) in the Mekong legal regime merely represents a crystallization of Thai-Vietnamese politics on a single print of time during the negotiation of the current treaty.

The downstream impacts in Cambodia of Vietnam's Yali Falls dam have been emblazoned on the international consciousness as a memorial to all that is wrong and pernicious with sectoral management of tributaries and unsustainable development. Consequences of the construction and operation of Yali Falls dam point to the inadequacy of Rule 5(A) for decision-making on tributary projects in the LMRB and underscore the significant transboundary effects and the legal imperative for reform. Article 5(A)'s aftermath proves that a state-centric approach to water resources development fails to provide satisfactory results on a regional scale and, ultimately, leads to a flashpoint signaling conflict. Vietnam's failure to notify Cambodia about development

1081 *Id.*

1082 Lebel and Garden, *supra* note 20, at 216.

1083 Thorson, *supra* note 1073, at 512.

and operations at Yali Falls had no consequences for the former state because the 1995 Mekong Agreement lacks viable enforcement mechanisms. The Yali Falls case study demonstrates that the LMRB countries, with creativity and resourcefulness, can utilize the MRC to the greatest extent possible for domestic development exercises without risking breach of the 1995 Mekong Agreement. Vietnam's manipulation of the "tributary rule" and the definition of "Proposed use" in the current treaty during the conduct of the Yali Falls project evidences that Article 5(A) facilitates each LMRB state in implementing megaprojects on tributaries by arguing sovereign rights while simultaneously ignoring sustainable water development, the presupposed overarching paradigm in the Mekong legal regime.

Dams on Mekong tributaries destroy the natural variable flow of rivers, impact aquatic species, flood valuable wetland and riverine habitat, prevent sediment movement downstream, and impede a river's natural ability to clean itself of harmful pollution. According to an International Rivers Network (IRN) report, "the consensus among river ecologists is that dams are the single greatest cause of the decline of river ecosystems."[1084] The report concluded that "[civil] society has come to understand dams can cause significant social and environmental impacts that outweigh the benefits they provide."[1085] Despite this knowledge, free-flowing tributaries throughout the LMRB are currently in danger of being dammed.

The grandiose Thai water grid reflects the disconnect between politically motivated water resources diversion projects and hydrologic reality and demonstrates a common theme in megaprojects on Mekong tributaries: the gap between the public and the policymakers. In its most ambitious form, the Thai water grid proposes to link the country's river basins by pipes and channels and alleviate natural allocation through sub-basin transfers (supplemented by interbasin transfers if necessary) from Mekong tributaries in Cambodia and the Lao PDR. These proposed transfers, which would effectively pipe water from the Lao PDR and Cambodia into Thailand, undoubtedly must be viewed suspiciously by stakeholders and community groups in the donor sub-basins as "water grabs" by the receiving sub-basins in Thailand. The grid illustrates the ease with which the LMRB states can circumvent the Mekong legal regime with collateral agreements due to the lack of interdependency in the 1995 Mekong

1084 IRN and American Rivers, "Beyond Dams: Options and Alternatives" (2004) 7; World Commission on Dams (WCD), "Dams and Development: A New Framework for Decision-Making" (WCD, 2000).

1085 *Id.* at 7.

Agreement. Undoubtedly, other options should be considered before piping water via intrabasin diversions across the LMRB.

In an attempt to overcome natural allocation and promote population growth and irrigation beyond the natural limits in the receiving sub-basins in Thailand, the proposed sub-basin transfers of the water grid will reduce the potential for long-term economic development and quality of life in donor sub-basins in the Lao PDR. Simply stated, this transfer stratagem moves water to where people are rather than moving people to where water is. These sub-basin diversions will interrupt river connectivity, disrupt fish spawning and migration, alter river geomorphology, and reduce stream flows in the donor basins, potentially producing adverse impacts on water quantity and quality (due to diminished assimilative capacity) and ecosystems. Trebling the irrigated land in Thailand will increase salinization in the already hypersaline receiving basins. The sub-basin transfers widely touted in the Thai water grid as a quick-fix solution to meeting escalating water demands in Isan and Bangkok are economically risky and come with significant environmental costs to the tributaries in the areas of origin. Large-scale water transfers from Mekong tributaries in the Lao PDR to Thailand will significantly alter the water balance in both the donor and receiving sub-basins.

The questions raised by these two projects are becoming increasingly important in the LMRB as many more projects are pending and proposed. The Yali Falls dam and Thai water grid case studies evidence the extent to which tributaries are marginalized in the Mekong legal regime and bear heavily on the overarching hydropolicy and legal issue in this research. What is clear from the case studies is that the LMRB states *cannot* simultaneously pursue hydro-sovereignty on tributaries and ensure the Mekong legal regime's efficacy to holistically cope with the emerging challenges and conflicts of transforming these states from developing to developed countries using the redefined river basin *approach* espoused in the 1995 Mekong Agreement. The disparate interests of the LMRB states that cause them to cling fervently to national sovereignty, and therefore to subscribe to Article 5(A), cannot be reconciled with the concept of the Mekong watercourse (basin) as a unitary whole. That provision of the treaty is directly at odds with the concept of a community of interest in the river and its tributaries. Although the 1995 Mekong Agreement codifies principles for regional cooperation, this oft lauded "agreement to agree" is for all practical purposes not applicable to the river's vital arteries and, therefore, sustainable development is a concept at the crossroads in the LMRB.

It seems there are not only legal, but also scientific imperatives for demarginalizing tributaries in the Mekong legal regime. The onslaught of dams built in violation of international law, such as Yali Falls, and questionable practices proposed in the water transfers of the Thai water grid evidence the need for regulation of water uses on Mekong tributaries. The uncertainty surrounding the Mekong tributaries' biogeography and the evolving contours of climate change law and policy raise relevant questions about the precautionary principle[1086] and principle of preventive action[1087]—two of the most effective elements of environmental protection of which the 1995 Mekong Agreement makes no mention. In the LMRB, arguably there is a need for more science to inform future decisions about water use in general and planned measures on tributaries in particular. In this respect, sound science can truly inform policy and result in better law in the Mekong legal regime. But if international law is to make a difference in holistic management of water resources and basin-wide governance in the LMRB, Article 5(A) concerning tributaries requires a fundamental rethink.

1086 Sands, *supra* note 616, at 267–268. "The precautionary principle aims to provide guidance in the development and application of international environmental law where there is scientific uncertainty .... On the one hand, some consider that it provides the basis for early international legal action to address highly threatening environmental issues such as ozone depletion and climate change. On the other hand, its opponents have decried the potential which the principle has for over-regulation and limiting human activity. The core of the principle, which is still evolving, is reflected in Principle 15 of the Rio Declaration, which provides that: Where there are threats of serious or irreversible damage, lack of full scientific certainty shall not be used as a reason for postponing cost-effective measures to prevent environmental degradation." *See generally,* Sands, *supra* note 616, at 266–279.

1087 Sands, *supra* note 616, at 247. "Broadly stated, ... [the principle of preventive action] prohibits activity which causes or may cause damage to the environment in violation of the standards established under the rules of international law." *See generally,* Sands, *supra* note 616, at 246–249.

CHAPTER 5

The Mekong: The Archetypal Dissevered River Basin

5.1 Introduction: The Dissevered River Basin Concept

The marginalization of tributaries and truncation of headwaters in the Mekong legal regime creates a basin that legally and on paper is in effect separated, divided into parts, and broken up. Based on analyses of tributaries in the Mekong legal regime and other selected international and interstate legal regimes in Chapter 3 and case studies in Chapter 4 of this book, I propose a new term in the hydropolicy and legal literature; a term that describes a generalized concept of a basin or river system in which tributaries and/or headwaters have been marginalized or excluded within any legal agreement or framework that redefines the basin, yet still uses the basin concept to create the legal fiction of a holistic water resources management paradigm in that legal agreement or framework: the *dissevered river basin*. The Mekong River Basin serves as an archetype of this concept, with its "marginalized" and "pruned" tributaries and "lopped" headwaters."[1088] The 1995 Mekong Agreement allows for the redefinition of the term basin by marginalizing tributaries and limiting the treaty's scope to the LMRB. The dissevered Mekong basin is emblematic of the disconnect between the notion of sovereignty and the physical hydrology of a river that transcends geopolitical boundaries. The LMRB states have adopted a "Westphalian" interpretation of sovereignty to develop tributaries unilaterally and sever hydro-interdependency, in effect creating four separate hydrologic units. Because of semantics, there is legally no Mekong "basin" but only a vestige of what that basin should be in terms of holistic water resources management manifested in the form of the redefined LMRB. Similarly, the legal regimes in the 1909 Boundary Waters Treaty, 1961 Columbia River Treaty, and 1928 Boulder Canyon Project Act all evidence the prevalence of the dissevered river basin concept in international treaties and interstate acts. The 1909 Boundary Waters Treaty excludes tributaries. The Okanagan River is not included in the 1961 Columbia River Basin Treaty. The 1928 Boulder Canyon Project Act reserved tributary watercourses in the Lower Colorado River Basin to the individual states. Indeed, the Colorado, like the Mekong, is divided into

1088 Carrard, *supra* note 10; Lebel and Garden, *supra* note 20, at 216.

 | DOI 10.1163/9789004362604_007

upper and lower basins using an arbitrary point, the former at Lee Ferry, the latter at the Chinese border. The Upper Colorado River Basin states address water allocation on tributaries in sub-compacts, while tributaries in the Lower Basin are reserved for state control and are not subject to interstate allocation. Each of these agreements and acts pretermits tributaries from their oft lauded legal regime, thereby creating dissevered river basins that foster piecemeal governance and prevent holistic water resources management. Dissevered river basins further hydrosovereignty and invite sectoral management of watersheds.

The 1995 Mekong Agreement uses the term "basin" throughout, but it fails to define the term, what it encompasses and how it should be interpreted. Additionally, in a flurry and blizzard of other undefined terms—"river basin," "basin level," "basin-wide" and "river system"—the text redefines the basin concept, seemingly, at every turn. In this figurative "basin," the headwaters are not included because of China and Myanmar's absence from the treaty. This leaves the skeletal remnant that is the LMRB, in which the MRC is only responsible for the governance of the mainstream. Tributaries are managed by each state through application of the 1995 Mekong Agreement's limiting tributary rule (Article 5(A)) which emphasizes hydrosovereignty. The tributary rule requires mere notification to the MRC JC for any intrabasin use and interbasin diversions and does not oblige any formal review, deliberation, or prior consultation among LMRB states before undertaking planned measures or megaprojects on tributaries. Further exacerbating the marginalization of tributaries is the definition of "Proposed use" in Chapter II of the 1995 Mekong Agreement, which excludes domestic and minor water uses that do not have a significant impact on mainstream flows. Accordingly, dams and water diversions on tributaries escape any meaningful oversight while dry season mainstream projects require consensus, irrespective of tributary projects' capacity to negatively impact the water quantity and quality downstream. Because several tributaries, such as the Sesan, are transboundary rivers, it is inconsistent to treat these international watercourses disparately from the mainstream. Indeed, the term "tributary" is not defined in the 1995 Mekong Agreement but rather is left as an afterthought in mere procedures (PNPCA) that are external to the treaty, and in all probability, are not legally binding. The PNPCA defines tributary as "a natural stream of the Mekong River System whose flows have a *significant* impact on the mainstream."[1089] Contributing further to the ambiguity of the PNPCA, the term "significant" is undefined.

1089 Procedures for Notification, Prior Consultation and Agreement, *supra* note 345, at sec. 1, para. 3 (emphasis added).

The modern trend and badge of responsibility for the LMRB states under the UN Watercourses Convention, which entered into force in August 2014, are to include an entire river basin, defined as including its tributaries, in international watercourse agreements. The limited extent to which planned measures and megaprojects on transboundary tributaries are subject to interdependency and basin-wide governance, remains problematic in the Mekong legal regime. Rules governing transboundary tributaries should comport with those for the mainstream in the 1995 Mekong Agreement.

Article 7 of the 1995 Mekong Agreement aptly demonstrates the shift in semantics used in the treaty to avoid embracing a unified drainage basin concept. Article 7 provides that the LMRB states are:

> To make every effort to avoid, minimize and mitigate harmful effects that might occur to the environment, especially the water quantity and quality, the aquatic (eco-system) conditions, and ecological balance of the *river system*, from the development and use of the Mekong River *Basin* water resources or discharge of wastes and return flows. Where one or more States is notified with proper and valid evidence that it is causing substantial damage to one or more riparians from the use of and/or discharge to water of the Mekong *River*, that State or States shall cease immediately the alleged cause of harm until such cause of harm is determined in accordance with Article 8. (emphasis added).

By using no fewer than three terms, river system, basin, and river, the treaty subtly redefines the basin for convenience. Coupled with the tributary rule, these undefined and ambiguous terms allow "tributary based projects ... [to] go ahead even when they have transboundary implications."[1090] Seemingly, there was a deliberate effort to leave states with a wide measure of control of tributaries and make the language ambiguous enough (i.e., constructive ambiguity) so that it would command general acceptance.

The redefinition of the term "basin" is also readily apparent in Article 35. If the MRC is unable to resolve a conflict, there is a subtle shift in words used in Article 35 that allows them "to proceed according to the principles of international law." This is probably a veiled reference to compulsory dispute settlement, to which the LMRB states would have to consent. Seemingly, this advocates a shift from a basin approach to that of an international watercourse. Accordingly, the Mekong "Basin" is constantly being redefined in the 1995

1090 Lebel et al. (2005), *supra* note 348, at 8.

Mekong Agreement, creating a dissevered river basin in the legal regime when convenient for the LMRB states' development plans and national agendas.

Article 38 of the 1995 Mekong Agreement in effect sanctions bilateralism in the river's legal regime and therefore promotes the dissevered river basin concept. Article 38 allows the parties to "[e]nter into bi- or multi-lateral special agreements or arrangements for implementation and management of any programs and projects to be undertaken within the framework of this Agreement, which agreements shall not be in conflict with this Agreement." China and Thailand have habitually endorsed bilateralism because, as Lebel et al. (2008), point out, they are "both willing and able to develop water resources on their own territory and make bilateral deals to meet their own rapidly expanding water, energy and transport demands."[1091] This strategy allows these two hegemons more bargaining power vis-à-vis one state rather than several. The Mekong riparians preference for bilateralism is also evident in the Thai water grid's rescaling with its proposed grandiose sub-basin and interbasin transfers,[1092] and the now defunct Quadripartite Economic Cooperation—China, Lao PDR, Myanmar, and Thailand's failed attempt under the 2001 Agreement on Commercial Navigation on the Mekong to establish commercial navigation on the upper Mekong.[1093] Needless to say, China is still doing its compassionate best to further navigation through "informal consultations."[1094] Effectively, then, the Mekong legal regime is hamstrung by its own mandate.

Some commentators posit that modern legal regimes must look beyond the basin concept to the politics of scale in water governance.[1095] The concept of the dissevered river basin lends some support to this position. Lebel et al. (2008) concisely and cogently summarize the limitations imposed by basin-oriented water management:

> Moreover, the hydrological unity of basins defined at a particular spatial level, are increasingly undermined by interbasin transfers and use of groundwater. There is no single, correct, area for managing water. Water politics is multiscale and invariably cuts across "basin" boundaries.
> ....

1091 Lebel et al. (2008), *supra* note 657, at 3.

1092 Lebel et al. (2005), *supra* note 348, at 7.

1093 *See* Oliver Hensengerth, "Transboundary River Cooperation and the Regional Public Good: The Case of the Mekong River" (2009) 31(2) *Contemporary Southeast Asia* 326, 334–335.

1094 Barry Wain, "Beijing's Clumsy Manoeuvres", *Far Eastern Economic Review* (26 August 2004); Hensengerth, *supra* note 1093, at 335.

1095 Lebel et al. (2008), *supra* note 657, at 1.

> The river basin is widely portrayed as the "natural" and "ideal" unit of management for water resources. There are some good reasons for this: the interconnections along waterways are important for fish and other aquatic organisms; many of the strong influences of land-use on river flows are within basin[s]; storage, diversion and use of water upstream affects water available or flood risks downstream. But there are also important limitations: basins are increasingly linked by physical diversions and transfers; groundwater interactions may not fully reflect surface flows; administrative and other political boundaries often do not correspond to basins; at levels below that of the entire basin, subdivisions are usually highly arbitrary.
>
> ....
>
> [T]he basin as it is currently institutionalized under the Mekong Agreement is a superficial, truncated, and lopped vision of a "basin". Key provisions of the Mekong Agreement apply to the mainstem of the river and not tributaries and only to the four downstream riparian countries.[1096]

Warner, et al., maintain that "[t]he choice to manage water on the basis of river basins is a *political* choice, and thus river basins are as much political units as they are natural units."[1097] The dissevered river basin in the Mekong legal regime is emblematic of such political units.

Beyond the basin concept, the two case studies, Yali Falls Dam and the Thai Water Grid, in Chapter 4 of this book, evidence the emergence of contested waterscapes that signal emerging issues of water allocation in the LMRB. The LMRB has continued to progress despite the pressures on water supplies from natural allocation and resulting overutilization of tributaries. This is primarily due to an abundant supply that is seemingly inexhaustible.

The 1995 Mekong Agreement was concluded in part to diminish the uncertainty in the inherent veto power in the previous legal regime. However, the current legal regime which marginalizes tributaries is passive. If holistic water resources management, sustainable development, and basin-wide governance are to be achieved in the LMRB, a true basin which includes tributaries, not just a redefined "paper basin" which marginalizes these important river arteries, must be adopted as a unit for water use and management. This could be readily accomplished by Cambodia, Lao PDR, and Thailand joining the UN Watercourses Convention, already ratified by Vietnam. In light of the projections of future water usage associated with tributary and mainstream

---

1096 *Id.* at 1–4.

1097 Jeroen Warner et al., *supra* note 574, at 121. (emphasis in original).

megaprojects, it is probable that some changes in international water law and policy will be necessary in the Mekong legal regime. As change dismantles the current legal regime, the question becomes: with what should the MRC replace it with? This question is considered in Chapter 7.

5.2 Chapter 5 Conclusion

The geographical and hydrological scope of the 1995 Mekong Agreement presupposes a basin framework for cooperation. However, what is clear in the final analysis of the Agreement is that parts of the Mekong Basin, such as tributaries and headwaters, are insufficiently covered in its text. Accordingly, I have shown that the river basin approach is not applied in the Mekong legal regime. Tributaries in the LMRB are marginalized and left to the whims of the state-centric NMCs by Article 5(A), and China and Myanmar are not parties to the Agreement, thus excluding the headwaters and UMRB from joint management. The MRC, therefore, is only responsible for the management of the mainstream.

The concept of the dissevered river basin is introduced to describe a basin in which tributaries and/or headwaters are marginalized or excluded within any legal agreement or framework that redefines the basin, yet still uses the basin concept to create the legal fiction of a holistic water resources management paradigm in that legal agreement or framework. A dissevered river basin creates separate hydrologic units by supplanting hydro-interdependency with hydro-sovereignty. This concept is useful as an analytical tool in reviewing international watercourse treaties for their capacity to manage water resources in a holistic manner with an ecosystem approach as described by McCaffrey and Weber.[1098] Generally, the more limited the extent to which planned measures on tributaries are subject to interdependent decision-making in an agreement, the more dissevered a basin is likely to be legally and on paper. The dissevered river basin concept is the hallmark of a legal regime constructed by the geopolitical boundaries of sovereignty at the expense of the basin's hydrology. In essence then, this concept severs hydro-interdependency. The 1957 Mekong Statute, Water Charter, and 1975 Joint Declaration adopted a basin approach (albeit limited to the LMRB) treating major tributaries the same as the mainstream. The current Mekong legal regime seemingly embraces the dissevered river basin concept supplanting joint cooperation with state sovereignty. The Yali Falls dam and Thai Water Grid case studies in Chapter 4 of this book

1098 McCaffrey and Weber, *supra* note 6, at 90.

demonstrate that, although intrabasin uses and interbasin diversions on tributaries are subject only to notification, these vital arteries, given their significant volumetric contributions to the Mekong, warrant further consideration in establishing instream and environmental flows on the mainstream.

Dissevered river basins are antithetic to the unitary whole of a watercourse and the community of interest concept supported by international case law (the *River Oder* case and *Gabcikovo-Nagymaros* case). International and interstate watercourse agreements and legislation are replete with examples of dissevered river basins (e.g., the 1909 Boundary Waters Treaty, 1961 Columbia River Treaty, and 1928 Boulder Canyon Project Act). Typically, these agreements and legislation marginalize tributaries and separate basins into redefined geopolitical units. Such practices have caused some commentators to question whether the basin approach is a viable concept for managing international watercourses when key provisions of legal regimes address only the mainstream, and marginalize or exclude tributaries and/or headwaters.

CHAPTER 6

A Flawed Vision? Deficiencies in the Mekong Legal Regime

The 1995 Mekong Agreement includes a framework for regional cooperation in the LMRB and is often lauded as a model of best practices for sustainable development in a transboundary watercourse.[1099] However, the current treaty has a number of deficiencies, in addition to—but nevertheless all interrelated to—the marginalization of tributaries, which include its lack of: (1) a basin-wide scope (China and Myanmar are not parties); (2) procedural elements of sustainable water development, namely EIAs, a legal framework for stakeholder participation and access to information; and (3) detailed dispute resolution provisions. These deficiencies are reviewed below.

6.1 Learning to Get Along: China and the LMRB States

Chinese dams in the Mekong cascade are already affecting the river's flow[1100] and will likely cause adverse transboundary environmental and economic impacts in violation of international law. As McCaffrey notes, "the Mekong case is an interesting one because it presents a combination of a modern treaty and a well (if externally) funded joint mechanism [the MRC], on the one hand and a regime of which the uppermost basin state and regional hegemon is not a member, on the other."[1101] From a Chinese perspective, self-exclusion from the Mekong legal regime is logical. As Hirsch sagaciously observes, "China has more to lose than gain from participation in a regime that restricts member states' power to act unilaterally."[1102] China's strategic upstream position, regional hegemony, "hydro-hegemony,"[1103] and preference for bilateral agreements explain that country's disinterest in joining what its leaders undoubtedly

1099 MRC, *supra* note 5.

1100 P. Khanna, *The Second World: Empires and Influence in the New Global Order* (Random House, 2008) 294–295.

1101 McCaffrey, *supra* note 9, at 288.

1102 P. Hirsch, "Beyond the Nation State: Natural Resource Conflict and 'National Interest' in Mekong Hydropower Development" (1999) 29 *Golden Gate Univ. L. Rev.* 399, 407.

1103 *See generally,* Zeitoun and Warner, *supra* note 332.

 | DOI 10.1163/9789004362604_008

view as a restrictive, multilateral legal regime. Moreover, China has focused considerable attention on its national development interests; specifically, large hydropower projects (the Mekong cascade) and removal of navigational obstacles (blasting rapids) on the upper Mekong, both of which are undoubtedly viewed by LMRB states as capable of causing "significant harm" to the river.[1104] China's stance on the 1995 Mekong Agreement, its vote against the 1997 UN Convention (Figure 2.1), and its unbridled dam development on the upper Mekong without notification to the LMRB states evidence an eschewing of multilateral negotiations with its downstream neighbors. The border skirmishes that led to the 1979 Sino-Vietnamese War serve as constant reminders of the potential for conflict between China and the LMRB countries.[1105] Considered in a vacuum, these facts alone may lead to somewhat of a misconception about the dynamics of Mekong hydropolitics.

China is often cited in the policy literature and popular press for its "territorial ambitions."[1106] However, noteworthy in Fravel's analysis is that China has settled 17 of its 23 territorial disputes through cooperative means, often with substantial compromises.[1107] Through a counterintuitive argument, Fravel attributes this fact to internal threats to regime security. China has participated as a "regular dialogue partner with the Mekong River Commission at yearly meetings since 1996."[1108] On 1 April 2002, China signed an agreement with the MRC to provide hydrological data to the LMRB states[1109] and renewed it on 29 August 2008.[1110]

The relationship between China and the LMRB states took a significant step forward toward basin-wide governance when, in November 2015, foreign ministers from all six Mekong nations launched the Lancang-Mekong Cooperation

1104 Lazarus et al., *supra* note 47.

1105 B. Elleman, "Sino-Soviet Relations and the February 1979 Sino-Vietnamese Conflict" (Draft Paper, 20 April 1996) *in* 1996 Vietnam Symposium, *After the Cold War: Reassessing Vietnam* (Texas Tech University, Lubbock, Texas, USA, 18–20 April 1996).

1106 M.T. Fravel, "Regime Insecurity and International Cooperation—Exploring China's Compromises in Territorial Disputes" (2005) 30 *International Security* 46; *See generally*, D. Roy, "Hegemon on the Horizon? China's Threat to East Asian Security" (1994) 19 *International Security* 149; G. Segal, *East Asia and the Constrainment of China*, 20 International Security 107 (1996).

1107 Fravel, *supra* note 1106, at 59.

1108 MRC, *supra* note 40, at 1.

1109 *Id.*

1110 MRC, "Mekong News" (Newsletter of the Mekong River Commission, July–October, 2008 (issue no. 3)) 6.

Mechanism (LMCM) during a meeting in Yunnan Province, China.[1111] This followed the 2010 release of dry season hydrological data regarding two of its dams located along the Lancang/Upper Mekong, which resulted after China received widespread criticism that its dam-building operations caused lower flows and decreased fish catches in the Lower Mekong.[1112] Although water resources is only one priority area over which the new partnership aims to enhance multilateral dialogue and cooperation, a particular declaration was issued in which the diplomats "agreed to 'enhance cooperation among [LMCM] countries in sustainable water resources management and utilization'".[1113] The cooperation has included the exchange of monitoring and water quality data, working together toward flood prevention and management, and sharing technical expertise concerning environmental management and hydroelectric projects.[1114] Notwithstanding the new cooperative spirit felt by the LMCM, it appears that China still remains relatively restrictive with sharing actual data from its Lancang dams, relying on the excuse that the data is linked to its national security interests.[1115] Although this is surely a positive development in multilateral governance, several concerns remain: when and in what circumstances China will release data (e.g., during emergencies); establishing a quantifiable and reliable point at which China will agree to release water from its dams during water shortages; and the possibility to implement a quasi-prior notification and/or consultation process between China and the MRC.

The partnership represents a continuation of a stronger Chinese role in the region, especially considering that Chinese companies bankrolled several of the other recently planned and completed hydropower projects in the LMRB. Also, as previously noted in Section 2.2.1 of this book, the MRC has no plenary jurisdiction and is not the only institutional framework for cooperation and partnerships in the LMRB. Indeed, the LMCM seeks to capitalize on multiple regional structures, to include the MRC, but also the Association of Southeast

1111 Sebastian Biba, *China Drives Water Cooperation with Mekong Countries* (2 January 2016) China Dialogue <https://www.chinadialogue.net/article/show/single/en/8577-China-drives-water-cooperation-with-Mekong-countries> (accessed 21 June 2017).

1112 *Id.*

1113 MRC, *Lancang-Mekong Cooperation: MRC Welcomes the New Initiative for Regional Cooperation by Six Countries in the Mekong River Basin* (Press Release, 31 March 2016) <http://www.MRCmekong.org/news-and-events/news/lancang-mekong-cooperation-MRC-welcomes-the-new-initiative-for-regional-cooperation-by-six-countries-in-the-mekong-river-basin/> [*hereinafter* LMCM Press Release] (accessed 19 September 2017).

1114 BDS 2016–20, *supra* note 367 at 34.

1115 Biba, *supra* note 1111.

Asian Nations Mekong Basin Development Cooperation and the GMS program.[1116] Consequently, if handled adroitly, these partnerships open new vistas for communication between China and the LMRB states in which the downstream countries can leverage navigation issues to promote transboundary governance in the Mekong.

Ultimately, it will be critical to involve China, within the general framework of the Mekong legal regime because of that country's control over Mekong headwaters. Seemingly, only when there is some political cost to China will its forward policy be constrained. However, it is doubtful that China can unilaterally exploit the Mekong on its own terms, largely because of the LMRB states' bargaining power over much broader issues such as navigation. Perhaps the most important issue impacting multilateral water governance between China and the LMRB states remains China's position on the UN Watercourses Convention. China was one of only three nations (along with Turkey and Burundi) to vote against adopting the convention when it was originally considered and adopted in the UN on 21 May 1997. At the time, the key criticism levied by the Chinese Permanent Representative to the UN was that language in the Convention institutionalized "the imbalance between rights and obligations of the upstream and downstream states"—presumably then, China wanted to protect its status as an upstream state and hydro-hegemon in the Mekong and other transboundary watersheds.[1117] In the years since, it does not appear that China has changed its position, instead choosing to negotiate bilateral agreements with individual nations regarding issues with a transboundary watercourse.[1118] Going forward, the ability to achieve basin-wide governance in the Mekong will be directly tied to whether China chooses to accede to the UN Watercourses Convention (admittedly unlikely, but possible nonetheless), updates or adopts bilateral treaties with particular principles and procedures embodied in the Convention, or maintains its current geopolitical strategy with the LMCM and other nonbinding institutions.

1116 LMCM Press Release, *supra* note 1113.

1117 Bhargav Sriganesh, *Compliance with UN Watercourses Convention: Half Full or Half Empty?* (unpublished dissertation, Kings College, London, 12 May 2017). *available at* http://www.e-ir.info/2017/05/12/compliance-with-un-watercourses-convention-half-full-or-half-empty/ (accessed 23 June 2017). *Citing* United Nations General Assembly Plenary-5-Press Release GA/9248 99th Meeting (AM) 21 May 1997. Statement by Gao Feng (China).

1118 *See generally,* Huiping Chen, Alistair Rieu-Clarke, and Patricia Wouters, "Exploring China's transboundary water treaty practice through the prism of the UN Watercourses Convention" (2013) 38 *Water Int'l* 217.

6.2 The View Downstream: EIAs

Article 7 of the 1995 Mekong Agreement provides that the member states are "to make every effort to avoid, minimize and mitigate harmful effects that might occur to the environment." The 1975 Joint Declaration required states to conduct environmental assessments of short- and long-term ecological impacts to other basin states before undertaking projects on the Mekong mainstream.[1119] Ironically, the 1995 Mekong Agreement, with its repeatedly touted sustainable development paradigm, does not legally require an EIA, if performed, to be provided to other watercourse states for any proposed development on the river. Article 12 of the UN Watercourses Convention provides that, before a planned measure is implemented, an EIA (if conducted), and the technical data and information for the project, shall be provided to states that may suffer a significant adverse effect. The 1995 Mekong Agreement was purportedly influenced by the ILC draft articles; hence, the lack of an EIA provision in the Mekong legal regime is a reasonable criticism. One of the key procedural elements of sustainable development, the EIA, then, is missing in the 1995 Mekong Agreement.

A glaring example of a substandard EIA was the original one performed for the Yali Falls project on the Sesan River. The inadequate EIA addressed effects of the project only 8 km downstream–solely within Vietnam. EVN initially neglected to consider any environmental impacts downstream in Cambodia. Essentially, the EIA fell below the minimum standard envisioned by Article 12 of the 1997 UN Convention. The lack of a legal requirement for transboundary EIAs is a significant deficiency of the 1995 Mekong Agreement especially with the incessant onslaught of megaprojects on transboundary tributaries.

The responsibility for promulgating guidelines for EIAs in the Mekong regime rests with the MRC's Environment Program.[1120] After more than 20 years since the 1995 Mekong Agreement's conclusion, in the aftermath of the disaster at Yali Falls, the treaty still contains no legal requirement for EIAs. Ostensibly, this evidences a lack of political will on the part of the member states to fully embrace sustainable development as the guiding principle in the treaty. In the *Case Concerning Pulp Mills on the River Uruguay* (*Argentina v. Uruguay*), the ICJ stated that the performance of an EIA for projects that may cause transboundary harm is now part of general (customary) international law.[1121] Some

1119 1975 Joint Declaration, *supra* note 246, art. XVII.

1120 MRC, *supra* note 895, at 16–17.

1121 Judgment of 20 April 2010, para.204, available on the ICJ web site, <http://www.icj-cij.org/docket/files/135/15877.pdf> (accessed 21 June 2017).

international legal regimes require environmental assessment procedures. For example, the 1991 European Convention on Environmental Impact Assessment in the Transboundary Context (Espoo Convention),[1122] which has the objective of preventing or reducing adverse transboundary impacts of proposed activities,[1123] can provide a guiding framework for the LMRB states to fully incorporate EIAs into the Mekong legal regime.

6.3 The Transparency Issue: Stakeholder Participation and Access to Information

The provisions of the 1995 Mekong Agreement make no reference to public participation or access to information, two interrelated key procedural elements of sustainable development and public policy development. In 1999, the MRC JC completed a report that provides certain responsibilities for member states regarding stakeholder participation. This report asserts that "directly affected people" are to have access to information and be involved in some level of decision-making.[1124] Although this report is a step in the right direction, it is generally non-binding on the MRC member states. Consequently, there is no legal framework for stakeholder participation as a legitimate process within the Mekong legal regime.

There was no stakeholder participation or access to information before the construction and operation of the Yali Falls dam on the Sesan River. Indeed, the affected communities were unaware that the dam was being constructed. Worse, in the aftermath of Yali Falls, the Vietnamese and Cambodian governments and the MRC did not immediately involve stakeholders in an active role in determining the effects of the dams on communities along the Sesan River.[1125] Further, no technical data on river fluctuations resulting from dam construction and operations or EIAs were provided to those affected as is envisioned by Article 12 of the UN Watercourses Convention. Likewise, the Thai water grid is shrouded in secrecy with little stakeholder involvement. If the 1995 Mekong Agreement is to function as a guiding framework for sustainable

1122 25 February 1991, 1989 U.N.T.S. 310 (entered into force 10 September 1997).

1123 *See generally*, A. Tesli and S.R. Husby, "EIA in a Transboundary Context: Principles and Challenges for a Coordinated Nordic Application of the Espoo Convention" (1999) 19 *Environmental Impact Assessment Review* 57.

1124 MRC, *Public Participation in the Context of the MRC* <http://www.MRCmekong.org/assets/Publications/policies/Public-Participation-in-MRC-context.pdf > (accessed 21 June 2017).

1125 Hirsch and Wyatt, *supra* note 802, at 63.

development and public policy development, adoption of a forward-thinking approach about transparency and public participation is fundamental. Such an approach is embodied in the 1998 Convention on Access to Information, Public Participation in Decision Making and Access to Justice in Environmental Matters (the Aarhus Convention).[1126]

6.4 Crossing the Rubicon: Dispute Resolution

The mechanisms for dispute resolution in the 1995 Mekong Agreement are considerably different from those specified in the UN Watercourses Convention. The UN Watercourses Convention provides detailed steps for the settlement of disputes which emphasize fact-finding, crisis management and creative dispute resolution.[1127] In contrast, the 1995 Mekong Agreement is vague and incomplete on what steps should be taken to provide remedies. Article 34 of the 1995 Mekong Agreement specifies the MRC as the primary institutional mechanism for dispute resolution and requires the MRC to "make every effort to resolve the issue." If the MRC is unable to remedy the dispute, the next step is for "the Governments to take cognizance of the matter for resolution by negotiation through diplomatic channels within a timely manner."[1128] Finally, Article 35 of the 1995 Mekong Agreement provides that "should the Governments find it necessary or beneficial ... they may ... request the assistance of mediation through an entity or party mutually agreed upon, and thereafter ... proceed according to the principles of international law."

The role of the relevant national institutions in resolving disputes should be considered in the Mekong water resources spectrum. The institutional, jurisdictional and legal complexity associated with water resources governance in the LMRB is indeed a significant barrier to dispute resolution, largely because, as Hirsch and Jensen point out:

> There is no real consistency between the [1995] Mekong Agreement and the national water policies and legislation of the States [NMCs] that are party to it. Moreover, scarcely any reference is made in the legislation and published policies of the participating countries to the [1995 Mekong]

1126 *See generally,* E. Petkova, and P. Veit, "Environmental Accountability Beyond the Nation-State: The Implications of the Aarhus Convention" (Environmental Governance Note, World Resources Institute, 2000).

1127 UN Watercourses Convention, *supra* note 18, art. 33.

1128 1995 Mekong Agreement, *supra* note 1, art. 35.

> Agreement or to any work of the MRC. If a part of the member states' mandate is to implement the MRC's programs, and to do so through their … [NMCs], there needs to be more consistency between jurisdictions.
>
> ….
>
> From a legal and institutional point of view, the MRC has some substantial obstacles to overcome before it can effectively use its authority to prevent the basin's resources being over-exploited and to encourage better conservation in the region. *Legal reform* is necessary to ensure that the high-level work now being undertaken and underwritten by the donor community is properly supported by tighter legal provisions at both the regional and national levels.
>
> ….
>
> The function of the NMCs also needs to be reconsidered. Ideally, the ministries represented in the NMCs should be taking back to their respective jurisdictions a coherent [and holistic] water resource policy framework, based on MRC guidelines that can be implemented. In the medium term, this ought to be translated into coherent legislative frameworks for [holistic] water resources management at [the] national level. It may well be that the "missing link" between MRC policy and its implementation at the national level is in fact the lack of a coherent basin-wide approach taken by the NMCs. This deserves further investigation.[1129]

In the LMRB states, several ministries share authority in a bewilderingly complex administrative galaxy that fails to address the underlying disputes. For example, in Cambodia, the Ministry of Environment and the Ministry of Water Resources and Meteorology share institutional authority for water management while in Vietnam the Ministry of Natural Resources and the Environment and Ministry of Agriculture and Rural Development attempt to balance water quality and water quantity, respectively, with overlapping jurisdictions that obviously should be integrated. There is scant policy in place to ensure water quantity, water quality, and ecosystem health are considered thoroughly in the dispute resolution process. In addition, internal agency processes and guidelines often do not provide clarity, transparency, and authority to resolve disputes. Holistic water resources management is hampered by state agencies'

1129 Hirsch and Jensen, *supra* note 180, at xvi and 42–44. (emphasis added).

responsibilities that overlap jurisdictions in environment, water, agriculture, and other natural resources departments. This evidences the overall need for comprehensive reform of water resources governance in the LMRB in terms of agency coordination and enhancement. Indeed, no single state agency or authority has oversight for ensuring the health and abundance of Mekong water resources. Instead, in the LMRB the task is spread out among multiple state agencies and the NMCs, often creating confusion, conflict, and inaction. Any solution in terms of dispute resolution should ensure transparency, allow public access to information, and involve local stakeholder and community groups, all nested within a basin-wide framework and policy.

The development of the Sesan River is again illustrative here. As noted in Section 4.1.2 of this book, to address the impacts of destructive flood pulses from construction and development of the Yali Falls dam, including estimated damages of US $3.3 million, 39 deaths and displacement of riparian villagers, the MRC facilitated establishment of the Cambodia-Vietnam Joint Committee for the Management of the Sesan River which met three times from 2000–2004, disbanded in 2004 and was reconstituted as a Standing Committee later that same year.[1130] The transition of responsibility to the Standing Committee (officially, the Cambodian Standing Committee for Coordination of Dams and Canals along the Cambodian-Laos-Vietnam-Thailand Borders) was precipitated primarily due to budgetary constraints, and the scope of the committee's responsibility was expanded to include all border dams, not just those in the 3S Basin.[1131] Still, the Cambodian Government continued to defund the initiative, which was evidenced in 2008, when a meeting between the Standing Committee and the counterpart Vietnamese committee was only made possible via funding from Vietnam, because the Standing Committee had no operating budget.[1132] It is unclear at this point—almost a decade later—whether the Standing Committee still engages in any activities whatsoever. Given their own national government's clear absence of a commitment to pursuing justice for the victims of the Yali Falls fallout, the likelihood that they will ever receive

1130 Michael Lerner, "Dangerous Waters: Violations of International Law and Hydropower Development along the Sesan River" (Background Briefing 4, National Sesan River Protection Network Workshop, 27 November 2002) 1–2; Wyatt and Baird, *supra* note 887, at 438.

1131 Ly Thim, *Planning the Lower Mekong Basin: Social Intervention on the Se San River* (Transaction Publishers, 2010) 125, 153.

1132 *Id. See also,* NGO Forum on Cambodia, "Down River: The Consequences of Vietnam's Se San River Dams on Life in Cambodia and Their Meaning in International Law" (prepared by Eric Rutkow, Cori Crider, and Tyler Giannini, December 2005) 8. *Available at* http://hrp.law.harvard.edu/wp-content/uploads/2013/02/Down-River-2005.pdf (accessed 19 September 2017).

reparations is slim at best. The MRC, Cambodia-Vietnam Joint Committee for the Management of the Sesan River and its successor, the Standing Committee, have yet to deal with the compensation issues or provincial requests for a moratorium on dam construction. Cambodia and Vietnam jointly requested assistance from the Asian Development Bank to strengthen arrangements for cross-border collaboration on the Sesan River. However, in the shadow of the Yali Falls dam aftermath, unmitigated impacts continue, and villagers remain uncompensated for their losses. All this time, Vietnam continued to develop dams on the Sesan River, and now Cambodia has plans in the works to construct a cascade of its own in the Lower Sesan and Sre Pok rivers. These convoluted, protracted procedures evidence the need for amended dispute resolution provisions in the Mekong legal regime. Again, the UN Watercourses Convention's dispute resolution procedures are instructive as a viable approach to strengthening the dispute settlement procedures in the 1995 Mekong Agreement.

6.5 The Mekong Legal Regime Viewed through the Lens of the UN Watercourses Convention

The deficiencies in the Mekong legal regime are perhaps most readily observed through the lens of the substantive and procedural frameworks of the UN Watercourses Convention which entered into force on 17 August 2014. According to Browder and Ortolano (2000), the 1995 Mekong Agreement is more comprehensive than the UN Watercourses Convention because:

> The Convention does not require basin states to establish a management organization, nor does it require a basin state to notify other basin states of proposed water use in all situations, nor does it require the basin [state] to cooperate for mutual benefit. The minimum standard of behavior in the convention is to try to avoid harm to other basin states and when harm may or does occur to negotiate a mutually acceptable solution.[1133]

It is asserted here that, when the two treaties are read *in pari materia*, the 1995 Mekong Agreement conspicuously lacks a number of basic principles and provisions included in the UN Watercourses Convention and is less robust. This substantially diminishes the former treaty's effectiveness in implementing holistic water resources management on Mekong tributaries, let alone

1133 Browder and Ortolano, *supra* note 241, at 527.

sustainable development. But considered together, the two treaties are generally compatible. As previously stated, the 1995 Mekong Agreement was drafted with the full benefit of the ILC draft articles of the UN Watercourses Convention. However, there are a number of provisions in the UN Watercourses Convention which have no counterparts in the 1995 Mekong Agreement. The main differences are:

- The UN Watercourses Convention defines "watercourse" to include both surface waters (including tributaries) and connected groundwater (Article 2(a)) and uses the terms "system" rather than "basin" terminology. The UN Watercourses Convention, then, incorporates a holistic and conjunctive approach to freshwater management or conjunctive management of surface water and groundwater. Further, the UN Watercourses Convention requires the states to define the waters that are the subject of an agreement (Article 3(4)). The 1995 Mekong Agreement refers to the Mekong "basin," as well as the terms "river basin," "basin level," basin-wide" and "river system" throughout, but the treaty fails to define those terms or the term "tributary," which is defined only in external "procedures" that are not legally binding on the LMRB states. The 1995 Mekong Agreement does not explicitly include tributaries or groundwater under binding provisions of the treaty. Thus, the UN Watercourses Convention approach is broader and could strengthen the 1995 Mekong Agreement by effectively demarginalizing tributaries and providing a true basin-wide conjunctive management framework which includes tributaries and groundwater.
- The UN Watercourses Convention obliges states to provide the results of EIAs, if conducted, for all planned measures which may have a significant adverse effect upon other watercourse states (Article 12). The Mekong legal regime does not legally require EIAs. Again, the UN Watercourses Convention can strengthen the 1995 Mekong Agreement by providing one of the procedural elements of sustainable water development: the EIA.
- The UN Watercourses Convention includes factors relevant to equitable and reasonable utilization (Articles 5–6) but the Mekong legal regime does not. In Article 6, the UN Watercourses Convention lists factors that should be considered when determining whether utilization is equitable and reasonable (e.g., a watercourse's natural characteristics, socio-economic needs of the parties, population dependent on the water in question, water conservation, vital human needs, and existing future uses). Thus, the former can inform the latter and could support the more general provisions of the 1995 Mekong Agreement regarding implementation of equitable and reasonable utilization, particularly human consumption.

- The UN Watercourses Convention includes an obligation not to cause significant harm (Article 7). The 1995 Mekong Agreement does not address significant harm, but rather Article 7 refers to "harmful effects" and requires the victim state to provide evidence of "substantial damage," both undefined terms in the current legal regime. The UN Watercourses Convention can assist the Mekong states in providing a nexus between the obligation of equitable and reasonable utilization and the stipulation to prevent significant harm.
- The UN Watercourses Convention does not differentiate between interbasin and intrabasin diversions, nor does it differentiate regarding exchange of information and notification of planned measures related to such diversions (Articles 5 and 11–14). Article 5 of the 1995 Mekong Agreement has a complex, convoluted matrix for reasonable and equitable use, although water is not allocated per se. This water use matrix emphatically marginalizes tributaries, treating them disparately compared with the mainstream.[1134] If all the LMRB states would join the UN Watercourses Convention, tributaries could be effectively demarginalized.
- The UN Watercourses Convention includes an obligation for protection of water quality and preservation of ecosystems, control of alien species and prevention, reduction, and control of pollution (Articles 7, 20–22). In Article 3 of the 1995 Mekong Agreement, the LMRB states merely agree to protect the environment, natural resources, aquatic life, and conditions and ecological balance of the basin from pollution or other harmful effects that stem from development plans and uses of water and related resources. As previously noted, Article 7 refers to harmful effects and substantial damage, two undefined terms in the Mekong legal regime. The UN Watercourses Convention *requires* joint efforts toward pollution control and mitigation, protection of marine ecosystems, and prevention of harmful conditions. The author is at a loss as to how the ecological balance of a large river can be maintained under a legal regime which lacks legally binding requirements to protect flow variability on tributaries. The answer, evidently, is that it cannot.
- The UN Watercourses Convention requires parties to negotiate in good faith and pay reasonable regard to the rights and legitimate interests of other states (Article 17 (2)). The 1995 Mekong Agreement contains no provisions pertaining to other states' interests that are not party to a particular negotiation. The Yali Falls dam case study, highlighted in Chapter 4 of this book, irrefutably supplies the best evidence for the need of such requirements, which are conspicuously missing in the Mekong legal regime.

1134 1995 Mekong Agreement, *supra* note 1, art. 5(A).

- The UN Watercourses Convention requires prevention and mitigation of harmful conditions, whether resulting from natural causes or human conduct, such as floods, ice conditions, erosion, siltation, drought, water-borne diseases, desertification, or saltwater intrusion (Article 27) and further, unlike the 1995 Mekong Agreement, provides for emergency situations (Article 28). There are no specific harmful conditions mentioned in the Mekong legal regime. The UN Convention could strengthen the Mekong legal regime by requiring expeditious notifications of impending threats to other watercourse states that occur from natural causes within their borders, compelling all states to regularly exchange information pertaining to emergencies, and requiring states to develop contingency plans for watercourse emergencies.
- Perhaps most importantly, the UN Watercourses Convention includes detailed dispute settlement provisions for constructive resolution of emerging environmental challenges at the intersection of law and policy, emphasizing strategic counseling, crisis management, and creative dispute resolution including the use of good offices, mediation, conciliation, arbitration, and reference to the ICJ (Article 33). The dispute resolution framework[1135] in the Mekong legal regime is convoluted and subject to protraction as illustrated by the Yali Falls dam case study.

These differences point to a broader scope, approach and framework in the UN Watercourses Convention in contrast to the 1995 Mekong Agreement. This does not auger well for future development on tributaries if Article 5(A) and other deficiencies of the Mekong legal regime continue to hold sway. The old regime of the 1957 Mekong Statute, the Water Charter, and the 1975 Joint Declaration reflected a broader approach similar to the UN Watercourses Convention. Vietnam ratified the UN Watercourses Convention in 2014 and Cambodia, Lao PDR, and Thailand should consider harmonizing and reconciling the current Mekong legal regime by also joining the treaty.[1136]

1135 1995 Mekong Agreement, *supra* note 1, arts. 34–35.

1136 UN Watercourses Convention, *supra* note 18, art. 3. Article 3 (2) of the UN Watercourses Convention calls on countries to "consider harmonizing such agreements [in force] with the basic principles of the present Convention." Further, Article 3(3) provides that "Watercourse States may enter into one or more agreements, hereinafter referred to as 'watercourse agreements', which apply and adjust the provisions of the present Convention to the characteristics and uses of a particular international watercourse or part thereof."

6.6 Chapter 6 Conclusion

After 20 years of disjunctive management in the Mekong River basin, it is evident that the current framework of the 1995 Mekong Agreement reflects a flawed vision. As demonstrated in this book, the Mekong legal regime lacks specific legal principles and mechanisms to regulate development on tributaries. In addition to marginalizing tributaries, it suffers from a lack of the following: (1) a basin-wide scope (the China issues); (2) procedural elements of sustainable water development (EIAs, stakeholder participation, and access to information); and (3) detailed dispute resolution provisions. In order to foster holistic water resources management and promote an ecosystem governance regime, the current treaty needs to be reconciled with the broader equitable framework of the UN Watercourses Convention. Cambodia, Lao PDR, and Thailand should join the UN Watercourses Convention. Otherwise, this large river system is destined for continued hydropolitical, economic, and environmental vulnerability.[1137]

1137 *See generally*, UNEP, *Hydropolitical Vulnerability and Resilience along International Waters: Asia* (UNEP, 2009) 4 ("Hydropolitical vulnerability" is defined by the risk of political dispute over shared water systems.).

CHAPTER 7

Conclusion: Demarginalizing Tributaries and Shifting the Mekong Legal and Policy Paradigm—A Proposal for the Way Ahead

The Mekong is the archetypal dissevered river basin with its tributaries managed and governed at the margin of the river's legal regime. Unilateral development of Mekong tributaries without adherence to the procedural requirements of sustainable development (EIAS, stakeholder participation, and access to information) is continuing at a disturbing rate in the LMRB, rendering the efficacy of the 1995 Mekong Agreement to support holistic water resources management and basin-wide river management questionable. This pursuit of hydrosovereignty portends the transformation of the LMRB. In addition, the Xayaburi dam planned by the Lao PDR on the Mekong mainstream illustrated the Achilles Heel of the 1995 Mekong Agreement, which evidently lacks effective mechanisms to prevent even the most brazen unilateral development of the mainstream.[1138]

"A disproportionately high amount of water" flows into the Mekong from tributaries including the Nan Ngum, Nam Theun, Sekong, Srepok, and Sesan rivers (Figure 1.3).[1139] This significant volumetric contribution is seasonal, largely results from flashy discharge (Figure 1.4) in the wet season, and is uneven along the mainstream.[1140] On average, tributaries provide 50 percent of the flow to the Mekong but, paradoxically, these arteries are rendered legally insignificant in the river's legal regime by Article 5(A), the "tributary rule."[1141]

Customary international law codified in the UN Watercourses Convention and the 1995 Mekong Agreement require notification for selected activities on Mekong tributaries which may have a significant effect downstream or constitute an emergency. Yet this is not enough. Tributaries should be demarginalized and brought from the shadows of the Mekong legal regime back

1138 For a discussion of the Xayaburi project on the Mekong mainstream and the surrounding controversy, *see* Alistair Rieu-Clarke, "Notification and Consultation Procedures Under the Mekong Agreement: Insights from the Xayaburi Controversy" (2014) 5 *Asian J. of Int'l L.* 143.

1139 Gupta, *supra* note 31, at 439–440.

1140 *Id.* at 439.

1141 Toda et al., *supra* note 188, at 5; Hirsch and Jensen, *supra* note 180, at 15; MRC, *supra* note 188, at 20.

 | DOI 10.1163/9789004362604_009

into the corpus of the 1995 Mekong Agreement. The Water Charter and 1975 Joint Declaration provided much more equitable and interdependent frameworks for governing tributaries and were indeed ahead of their time in the context of holistic water resources management. Continued disparate treatment of tributaries in the Mekong legal regime is an invitation for increased hydro-hegemony and a recipe for distrust and unresolved conflict, perhaps evidenced best by the aftermath of the Yali Falls disaster and continuing development of the Sesan River in Vietnam and Cambodia. McCaffrey's foretelling of future problems emanating from this disparate handling of tributaries in the Mekong legal regime is now somewhat prophetic.[1142] A key question is, can "contested waterscapes"[1143] be prevented on tributaries under the current legal regime in the LMRB? Irrespective of the disparate treatment of tributaries in the Mekong legal regime in Article 5(A), the LMRB states remain obligated under the UN Watercourses Convention, which codifies customary international law, and the general framework of the 1995 Mekong Agreement to cooperate (Article 1), protect the environment and ecological balance (Article 3), reasonably and equitably use water (Article 5), maintain flows on the mainstream (Article 6), and prevent transboundary harm (Article 7) in the Mekong River *Basin*, not just on the mainstem.

According to renowned geoscientist and conflict mediator Aaron Wolf, et al., "[i]t turns out that very rapid changes, either on the institutional side or in the physical system [such as large-scale dams on transboundary tributaries or water diversion megaprojects], which outpace the institutional capacity to absorb that change, are at the root of most water conflict."[1144] Based on historical incidents of international water cooperation and conflict from 1948–1999 in the International Water Events Database,[1145] the Mekong was classified in the "Basins at Risk" project[1146] as a basin with a setting that suggests the potential

---

1142 McCaffrey, *supra* note 9, at 287. In 2007, from his analysis of the Mekong legal regime, McCaffrey stated that "The 1995 Agreement requires only notification but not consent for projects on tributaries, which may give rise to problems in the future."

1143 *See generally,* Molle et al., *supra* note 572 (describing the concept of contested waterscapes).

1144 Aaron T. Wolf et al., "International Waters: Identifying Basins at Risk" (2003) 5(1) *Water Policy* 29.

1145 Oregon State University Institute for Water and Watersheds, *International Water Events Database: 1950–2008*, Available at <http://www.transboundarywaters.orst.edu/database/interwatereventdata.html> (accessed 23 June 2017).

1146 The "Basins at Risk" project is the brainchild of Aaron T. Wolf, Director, Program in Water Conflict Management and Transformation, Oregon State University, and US State Department policy analyst Shira B. Yoffe, who developed it to address "a series of overarching

for political stresses in the coming five to ten years.[1147] An update of events data from 2000–2008 in the International Water Events Database suggests a shift toward less cooperative relationships in the Mekong.[1148] Noteworthy is that when the events from 2000–2008 are broken into issues, infrastructure (such as the unbridled development of dams on Mekong tributaries) and water quantity are the catalyzing mechanisms in the majority of the events.[1149] In their analysis of these events, De Stefano et al. found that "[i]nfrastructure and water quantity seem to consistently be the aspects of trans-boundary water management most likely to precipitate conflict, and are indicating an increase in the weight of negative events in recent times."[1150] Wolf posits that the new data in the Mekong point to a basin "still 'at risk' precisely for the reasons" outlined in this book, namely tensions from development of infrastructure on Mekong tributaries and water quantity issues.[1151] Continued "unsustainable

---

gaps in research on freshwater resources and international conflict by providing a quantitative global scale exploration of the relationship between freshwater resources and conflict. The project incorporated a spatial perspective and considered the full spectrum of interactions, using precise definitions of conflict and cooperation. The purpose of the research was to identify historical indicators of international freshwater conflict and cooperation and from them, create a framework to identify and evaluate international river basins at potential risk for future freshwater conflict. Completion of the Basins at Risk project required three main elements: creation of an event database [the International Freshwater Event Database] documenting historical water relations, including a methodology for identifying and classifying events by their intensity of cooperation and conflict; construction of a geographic information system (GIS) of countries and international basins, both current and historical; and the collection or creation of indicator variables (biophysical, socioeconomic, and geopolitical) for testing of hypotheses about factors associated with water conflict." *See generally*, Wolf et al., *supra* note 1144; Shira B. Yoffe, Basins at Risk: Conflict and Cooperation Over International Freshwater Resources (2001) (unpublished Ph.D. dissertation, Oregon State University), available at <http://www.transboundarywaters.orst.edu/research/basins_at_risk/index.html> (accessed 23 June 2017).

1147 Wolf et al., *supra* note 1144, at 46–47.

1148 Lucia De Stefano et al., "Updating the International Water Events Database" (UNESCO Potential Conflict into Cooperation Potential (PCCP) Project, Dialogue Paper 7, 2009) 9. [hereinafter De Stefano et al. (2009)]; Lucia De Stefano et al., "Tracking cooperation and conflict in international basins: historic and recent trends" (2010) 12(6) *Water Policy* 871, 880. [hereinafter De Stefano et al. (2010)].

1149 De Stefano et al. (2009), *supra* note 1148, at 6; De Stefano et al. (2010), *supra* note 1148, at 877.

1150 De Stefano et al. (2009), *supra* note 1148, at 6.

1151 See email from Aaron T. Wolf, Director, Program in Water Conflict Management and Transformation, Oregon State University, to author (14 April 2010, 10:18 p.m. CST) (on file with author).

basin management"[1152] without regard to cooperative management and governance of Mekong tributaries will only add to these political stresses. The question then, becomes whether the LMRB can absorb the rapid changes currently taking place. Indeed, the United States reinvigorated their relationship with the Mekong countries in 2009 in order to address the myriad of environmental, hydrological, and geopolitical changes the basin is undergoing by establishing the Lower Mekong Initiative (LMI). The LMI is branded as a "multi-national partnership among Cambodia, Laos, Myanmar, Thailand, Vietnam, and the United States to create integrated sub-regional cooperation among the *five* Lower Mekong countries"[1153] (conspicuously, China is not included in the LMI framework or activities). The US Agency for International Development has taken a leading role in funding and implementing many of the program areas envisioned by the LMI, helping the Mekong countries with better management of their shared water resources in the following areas: technological cooperation, data standardization, and information sharing; improving EIAs and social impact assessments; stakeholder involvement in decision-making procedures; and enhancing the resilience of individual farmers and ordinary villagers by educating them about the impacts of climate change and how they can adapt their local economies accordingly.[1154] Despite the growing amount of regional and international partnerships available to the LMRB countries, the principle concern remains that the underlying legal regime cannot adequately support a sustainable development paradigm, rendering initiatives like the LMI and LMCM much less effective.

If the Mekong River basin is not managed properly, another era of conflict could develop. Sustaining the Mekong's tributaries is one of Southeast Asia's gravest social and environmental challenges that is emblematic of the inherent conflict between the LMRB states' desire for energy and irrigation and sustainable water development. Holistic water resources management and basin-wide governance are sorely needed to preserve the Mekong's fragile ecosystem, but sadly these do not appear to be on the horizon. However, Vietnam accepted these challenges when it ratified the UN Watercourses Convention in 2014. It is time for Cambodia, Lao PDR, and Thailand to follow Vietnam's leadership.

The 1995 Mekong Agreement lacks the substantive and procedural frameworks to promote basin-wide, long-term holistic water resources management. It is argued here that a more comprehensive legal regime such as those

---

1152 Rix, *supra* note 800, at 104.

1153 Lower Mekong Initiative, *The Lower Mekong Initiative* (*LMI*) <http://lowermekong.org/about/lower-mekong-initiative-lmi> (accessed 23 June 2017). (emphasis added)

1154 United States Agency for International Development, *Mekong, Mother of Water*, <https://www.usaid.gov/global-waters/june-2015/mekong> (accessed 23 June 2017).

embodied in the Water Charter, 1975 Joint Declaration, and UN Watercourses Convention is needed to manage the LMRB and to shift the paradigm of the Mekong legal regime toward sustainable development. In light of the most recent planned measures on the Mekong mainstream, seemingly unstoppable construction of dams on tributaries, and the Chinese "Lancang cascade," it is arguable that a new agreement would look substantially different.[1155] Changes are needed to provide a more precise legal mandate for the MRC and to foster holistic water resources management. Water allocation, flow regimes, and transboundary tributaries should not be left to external procedures which in effect relegate the 1995 Mekong Agreement to a "read between the lines" framework and are, almost certainly, not legally binding. Based on my analysis of tributary watercourse law in the Mekong legal regime, the most effective path forward for the LMRB states is coupling or packaging the UN Watercourses Convention with the 1995 Mekong Agreement. Vietnam has moved forward on this proposal by joining the UN Watercourses Convention. There are numerous benefits to the other LMRB countries by joining the UN Watercourses Convention summarized in Table 7.1.[1156]

Analysis of five decades of Mekong cooperation reveals that, although there has been considerable focus on a mainstream cascade, tributaries have been the sites of most megaprojects. The lack of substantive and procedural frameworks in the Mekong legal regime at the tributary or sub-basin level and lack of institutional capacity in the MRC increases the potential for transboundary water conflict in the LMRB. The LMRB states need to consider *all* aspects of the Mekong and its drainage system in their context. This requires looking at this basin from its headwaters and tributaries to its terminus in the South China Sea.

1155 For commentaries see, B.L. Bearden, "Following the Proper Channels: Tributaries in the Mekong Legal Regime", 14(6) *Water Policy* (2012) 991–1014 and B.L. Bearden, "The Legal Regime of the Mekong River: A Look Back and Some Proposals for the Way Ahead" (2010) 12(6) *Water Policy* 798.

1156 Sources: R. Kinna and A. Rieu-Clarke, "The Governance Regime of the Mekong River Basin" (2017) 2 *Int'l Water L.* 1; IUCN, "A Window of Opportunity for the Mekong Basin: The UN Watercourses Convention as a Basis for Cooperation (A Legal Analysis of How the UN Watercourses Convention Complements the Mekong Agreement)" (IUCN, 2016) 27; B. Bearden, A. Rieu-Clarke, and S. Pech, "Mekong Basin" in F. Rocha Loures and A. Rieu-Clarke (eds.), *The UN Watercourses Convention in Force—Strengthening International Law for Transboundary Water Management* (Routledge, 2013) 180–188; B.L. Bearden, "Following the Proper Channels: Tributaries in the Mekong Legal Regime" (2012) 14(6) *Water Policy* 991; B.L. Bearden, "The Legal Regime of the Mekong River: A Look Back and Some Proposals for the Way Ahead", 12 *Water Policy* (2010) 798–821.

TABLE 7.1 *Benefits to the Lower Mekong River Basin States Joining the UN Watercourses Convention*

Benefit	Parties that would benefit
Increased UNESCO resources (Tonle Sap)	Cambodia
Unchanged status as non-parties to the 1995 Mekong Agreement	Myanmar, China
Inclusion of tributaries and groundwater	All stakeholders
Could support the more general provisions of the 1995 Mekong Agreement regarding equitable and reasonable utilization, particularly human consumption	All stakeholders
Political feasibility versus renegotiating the 1995 Mekong Agreement	LMRB countries
Could help to provide a nexus between the obligation to equitable and reasonable utilization and the stipulation to prevent significant harm to the environment	All stakeholders
Strengthens the requirement for regular data and information exchange	All stakeholders
More open and transparent environment for development notification and negotiation process, data sharing, evaluations, and dispute resolution	Mainly the public, environmental NGOs, other stakeholders currently kept in the dark
More detailed, definitive, and independent dispute resolution process; could lead to more effective, timely, and equitable results	All stakeholders
Enhanced ability of private citizens to seek legal recourse for harm originating from another watercourse state	Individual stakeholders in Mekong Basin
Improves the overall governance of the Mekong Basin, while not challenging the sovereignty of any individual nation, nor having to sacrifice the "Mekong Spirit"	All stakeholders
Legal scope of the 1995 Mekong Agreement would be expanded	MRC's "basin-wide approach"
The 1995 Mekong Agreement would become aligned with customary international law	MRC, LMRB countries

TABLE 7.1 *Benefits to the Lower Mekong River Basin States Joining the UN Watercourses Convention* (cont.)

Benefit	Parties that would benefit
Would strengthen customary international water law in general	The broader international community.
Allows the states to become involved in the progressive development of the UN Watercourses Convention	All Mekong countries; broader international community indirectly
Increased willingness of China to engage in greater cooperation, involvement	LMRB countries
Affords the opportunity for individual nations to update their domestic laws to align with the UN Watercourses Convention and the 1995 Mekong Agreement	All Mekong countries
Potential economic benefits by enhancing initial data-sharing, notification and negotiation processes (versus more protracted negotiations or dispute resolutions)	All stakeholders

Mekong tributaries contribute half of the flow to the mainstream (Figure 1.3). This establishes the significance of Mekong tributaries in hydrological terms and therefore reinforces the dangers associated with failing to deal with them adequately in the political equation in terms of holistic water resources management. In the Mekong legal regime, strengthening international law would improve holistic water resources management and basin-wide governance. Accordingly, a legal fusion of sorts with the UN Watercourses Convention is required to bring dissevered tributaries and headwaters back from the margins of the 1995 Mekong Agreement into the core of the legal regime and demarginalize them. Currently in the Mekong legal regime, tributaries are reserved for use in national agendas.

Analyses of the complex and convoluted matrix in Article 5 and the other provisions of the 1995 Mekong Agreement beg the question of whether the Mekong legal regime is really about water sharing or serves some other agenda. This is not a difficult policy analysis to perform, largely due to the conspicuous lack of a number of the provisions of the UN Watercourses Convention in the 1995 Mekong Agreement; principally the Agreement's failure to define the term "basin," which allows for its continuous redefinition. Indeed, even when the current treaty is read charitably, there is very little substantive water law in the Mekong legal regime.

The four LMRB states should continue negotiations and maintain an ongoing open dialogue with China and Myanmar to increase opportunities for holistic water resources management and basin-wide governance, by including of these two states in the Mekong legal regime where there are issues of mutual interest such as navigation. There are legal and scientific imperatives to bring China into the legal regime because of its control of the Mekong headwaters. China has participated as a regular dialogue partner with the MRC since 1996 and has shared limited hydrological data with the LMRB states since 2002. China's participation is a step toward closer cooperation with the LMRB states.

If the LMRB states, with all their past and present tensions, can find areas of compromise, then surely the other two riparians can do the same. As McCaffrey suggests, "it seems possible that the well-known 'Mekong Spirit', which enabled the four Lower Mekong countries to continue cooperating even during wartime, will increasingly embrace the two upper riparians."[1157] "This," he notes, "together with a robust river basin organization, augers well for good relations among the Mekong countries, and improved management of the river, in the future."[1158]

Past hydropower development events on the Sesan River and proposed water diversions in the Thai water grid pointedly highlight the soft nature of the Mekong legal regime. Overall, the 1995 Mekong Agreement remains very much a hortatory, general framework with a relatively undefined role for the MRC in the Agreement itself. The 1995 Mekong Agreement represents a significant expansion in scope (sustainable development) over the previous Mekong agreements, but the 1995 treaty is in most other respects substantially weaker in its core provisions (Articles 3, 5, 6 and 7) than its predecessors in terms of interdependency, basin-wide governance and holistic water resources management.[1159] Merely including concepts in a framework instrument is not enough. The marginalization of tributaries in Article 5(A), the Yali Falls dam and Thai water grid case studies in Chapter 4, and the selected deficiencies in the 1995 Mekong Agreement highlighted in Chapter 6 support this summary conclusion and evidence a greater focus on state sovereignty than in the language found in the Water Charter, 1975 Joint Declaration, and the UN Watercourses Convention.

It is now widely accepted that holistic water resources management should be accomplished at the basin level.[1160] The river basin is a logical administrative

1157 McCaffrey, *supra* note 9, at 288.

1158 *Id.*

1159 Elhance, *supra* note 37, at 191–216; Nakayama, *supra* note 41, at 305–306.

1160 McCaffrey and Weber *supra* note 6, at 90.

unit to handle the regional externalities linked to conflicts of water use on transboundary tributaries. Legal regimes such as those embodied in the 1995 Mekong Agreement, 1909 Boundary Waters Treaty, 1961 Columbia River Treaty and Law of the [Colorado] River (Lower Basin), which marginalize or exclude tributaries, seemingly, are too often the norm in international and interstate practice today. Indeed, as Wolf observes, "While water managers generally understand and advocate the inherent powers of the concept of a … [basin] as a unit of management, where surface and groundwater quantity and quality are inexorably connected, the institutions which have developed to manage the resource have historically followed these tenets only in the exception."[1161] The basin is not an end point, but rather a point of beginning for holistic water resources management as well as sustainable water development. By examining all actions on tributaries concomitantly with those on the mainstream, states begin to move towards an ecosystem-based legal regime.

The 1995 Mekong Agreement is today one of the few legally binding multilateral treaties in Southeast Asia. While any agreement promoting cooperation, protection, and equitable utilization in an international river basin as important and undeveloped as the Mekong is certainly welcome, further work needs to be undertaken to pave the way ahead. Although the 1995 Mekong Agreement in its present form cannot be regarded as the most effective legal regime for holistic water resources management and sustainable development in the LMRB, it is nevertheless a very important first step in that direction. A weak treaty consisting largely of soft law, such as the 1995 Mekong Agreement, cannot counter China's regional hydro-hegemony in the UMRB or the interstate hydro-hegemony within the LMRB.[1162] In the LMRB, the continuing controversy concerning the victims of impacts from the construction and current operations of Yali Falls dam, and the seeming unwillingness of the Vietnamese and Cambodian governments to take the necessary steps to craft a just outcome for them, is dispositive of this. If the LMRB countries are to become fully secure and avoid further conflict, they need a much better strategy than hydrosovereignty. That strategy is holistic water resources management through regional cooperation, which can be realized by combining soft law and hard law found in the conceptual frameworks of both the 1995 Mekong

1161 Aaron T. Wolf, "The Present and Future of Transboundary Water Management" in C.M. Figueres et al. (eds.), *Rethinking Water Management: Innovative Approaches to Contemporary Issues* (Earthscan, 2003) 168.

1162 *See generally,* Melvin Woodhouse and Mark Zeitoun, "Hydro-hegemony and International Water Law: Grappling with the Gaps of Power and Law" (2008) 10(Supp. 2) *Water Policy* 103.

Agreement and the UN Watercourses Convention. Unless the LMRB states are willing to put back into the current Mekong legal regime the same considerable political will it took to negotiate the Water Charter and 1975 Joint Declaration and the holistic water resources management and ecosystem approach embodied in the UN Watercourses Convention, already ratified by Vietnam, then obviously the presupposition of sustainable development upon which the regime is based is bound to be fallible. As McCaffrey notes:

> It is ... [commonplace] today that per capita availability of fresh water is shrinking. [S]ome 60 per cent of global freshwater flows are contained in the 263 river basins that are shared by two or more countries, and that around 40 per cent of the human population lives in these international basins. These facts underscore the *necessity* of cooperation between states sharing fresh water, whether it is on the surface or underground. And yet internal political forces often lead countries to maximize their use of shared water resources without considering adequately the needs of their neighbors and co-riparians. Shrinking availability of water will only exacerbate this tendency, resulting in the potential for increased conflict. These factors demonstrate the importance of generally accepted legal norms governing the use by states of shared water resources. A set of such norms is contained in the 1997 United Nations Watercourses Convention. The Convention, which is largely a codification of customary international law, may be seen as an expression of the meaning of state sovereignty as it relates to transboundary water resources. Sovereignty over such resources while they are in a state's territory is not absolute. It is informed and inflected by the requirement that the needs and interests, expressed as rights, of co-riparians be taken into account. It is in this sense that a country's "sovereignty" over internationally shared water resources should be understood.[1163]

Perhaps it depends on the willingness of Cambodia, Lao PDR, and Thailand to follow the effective diplomacy of Vietnam in the Mekong legal regime by joining the UN Watercourses Convention, embracing sustainable development,

1163 Stephen C. McCaffrey, "Sovereign and Cooperative Management of Shared Water Resources in a Time of Shrinking Availability: The Role of International Law, Transboundary Aquifers: Challenges and New Directions" (Abstracts, UN Educational, Scientific and Cultural Organization (UNESCO) International Hydrological Programme (IHP), ISARM 2010 International Conference, Paris, 6–8 December 2010) 114. (emphasis added).

and following the "proper channels"[1164] of international water resources law and policy. But as Cambodian, Lao PDR, and Thai villagers living along the Mekong's vital arteries and those states' government officials know all too well, the "*proper channels* are [often] the ones full of strategic blockages and diversionary tributaries."[1165]

1164 David Edelstein, *Mr. Fiennes, I Presume?* (30 August 2005) Slate, available at < http://www.slate.com/id/2125293/?nav=fo > (accessed 23 June 2017).

1165 *Id.* (emphasis added).

Bibliography

Affeltranger, Bastien, "Intra-basin conflict resolution in the Mekong basin: Is a reconciliation of water values possible?" (International Workshop on the Value of Water—Different Approaches in Transboundary Water Management, Koblenz, Federal Republic of Germany, March 10–11, 2005).

Agenda 21, Report of the United Nations Conference on Environment and Development, U.N. Doc. A/CONF. 151/26/Rev. 1 (1993), Davison for Sustainable Development, UN Department for Economic and Social Affairs, available at <https://sustainabledevelopment.un.org/content/documents/Agenda21.pdf>, (accessed 8 June 2017).

Alexander, R.B. et al., "Effect of Stream Channel Size on the Delivery of Nitrogen to the Gulf of Mexico" (2000) 403 *Nature* 758.

American Bar Association, Cambodian Law and Democracy Project, "Needs Assessment Report," Phnom Penh Office (June 1995) available at <http://pdf.usaid.gov/pdf_docs/PNABX357.pdf> (accessed 14 June 2017).

Andersen, Inger, Ousman, Dione, Jarosewich-Holder, Martha, Olivry, Jean-Claude and Golitzen, Katherin G. (ed.), *The Niger River Basin: A Vision for Sustainable Management* (The World Bank, 2005).

Andrassy, Juraj, "Les Relations Internationales de Voisinage," (1951-II) 79 *Recueil Des Cours* 77–181.

Asian Development Bank, "Building a Sustainable Energy Future: The Greater Mekong Subregion" (GMS Report, 2009) 9.

Asian Development Bank, *GMS Sector Activities* <https://www.adb.org/countries/gms/sector-activities> (accessed 14 June 2017).

Backer, Ellen Bruzelius, "The Mekong River Commission: Does It Work, and How Does the Mekong Basin's Geography Influence Its Effectiveness?" (2007) 26(4) *Journal of Current Southeast Asian Affairs* 31.

Baird, Ian G. and Mean, Meach, *Sesan River Fisheries Monitoring in Ratanakiri Province Northeast Cambodia: Before and After the Construction of the Yali Falls Dam in the Central Highlands of Vietnam*, 3S Rivers Protection Network (3SPN) in cooperation with the Global Association for People and the Environment (GAPE) Ban Lung, Ratanakiri Province, Northeast Cambodia (December 2005).

Bakker, Karen, "The Politics of Hydropower: Developing the Mekong" (1999) 18 *Pol. Geo.* 209.

Bangkok Post, *Rules being drafted not fair to Thailand. Three-year talks still far from a deal* (11 November 2003).

Bangkok Post, *Water crisis looms, says grid study* (13 June 2004).

Bates, R.L. and Jackson, J.A. (eds.), *Glossary of Geology* (American Geological Institute, 1980).

Bearden, Bennett L., "The Legal Regime of the Mekong River: A Look Back and Some Proposals for the Way Ahead" (2010) 12(6) *Water Policy* 798.

Bearden, Bennett L., "Following the Proper Channels: Tributaries in the Mekong Legal Regime" (2012) 14(6) *Water Policy* 991.

Bearden, Bennett, Rieu-Clarke, Alistair and Pech, Sokhem, "Mekong Basin" in F. Rocha Loures and A. Rieu-Clarke (eds.), *The UN Watercourses Convention in Force—Strengthening International Law for Transboundary Water Management* (Routledge, 2013) 180–188.

Benda, Lee et al., "The Network Dynamics Hypothesis: How Channel Networks Structure Riverine Habitats" (2004a) 54 *Bioscience* 413.

Benda, Lee et al., "Confluence Effects in Rivers: Interactions of Basin Scale, Network Geometry, and Disturbance Regimes" (2004b) 40 *Water Resources Research* WO5402, doi:10.1029/2003WR002583.

Beyerlin, Ulrich, "The Concept of Sustainable Development" *in* R. Walfram (ed.), *Enforcing Environmental Standards: Economic Mechanisms as Viable Means?* (Springer, 1996) 95–121.

Biba, Sebastian, *China Drives Water Cooperation with Mekong Countries* (2 January 2016) China Dialogue <https://www.chinadialogue.net/article/show/single/en/8577-China-drives-water-cooperation-with-Mekong-countries> (accessed 21 June 2017).

Binnie and Partners, "Mun River Basin Water Resources Development Master Plan, Final Technical Report" (Prepared by Binnie and Partners in association with WS Atkins International Ltd., UK; ITC, Netherlands; ATT Consultants Co. Ltd. and TA&E Consultants) prepared for Royal Irrigation Department, Ministry of Agriculture, Bangkok, Thailand (1995).

Bird, Jeremy, "Introductory Speech" (Speech delivered as new MRC Secretariat CEO, Vientiane, Lao PDR, 7 April 2008).

Birnie, Patricia W. and Boyle, Alan E., *International Law and the Environment* (Oxford University Press, 2nd ed., 2003).

Bonheur, Neou, "National Report on the Prevention and Resolution of Environmental Conflicts in the Lower Mekong River Basin (Cambodia)," Prepared for Cambodian NMC and MRC (November 2002).

Browder, Greg, *Negotiating an International Regime for Water Allocation in the Mekong river basin* (unpublished Ph.D. dissertation, Stanford University, 1998).

Browder, Greg and Ortolano, Leonard, "The Evolution of an International Water Resources Management Regime in the Mekong River Basin" (2000) 40 *Nat. Resources J.* 499.

Brown, D.J. and Coon, T.G., "Abundance and Assemblage Structure of Fish Larvae in the Lower Missouri River and its Tributaries" (1994) 123 *Transactions of the American Fisheries Society* 718.

Bruce, J.P. et al., "Climate Change Impacts on Boundary and Transboundary Water Management" (Climate Change Action Fund Project Natural Resources Canada Project No. A458/402, 30 June 2003) 2.

Buxton, Michael et al., "Environmental Conflicts in the Mekong River Basin: Prevention and Resolution" (Report to the Environment Programme, Mekong River Commission, School of Social Science and Planning RMIT University, Melbourne, Australia 2003).

Buxton, Michael et al., "Conflict Resolution and Policy Making Mediation in the Mekong River Basin" (September 2006) 41 *Just Policy* 26.

Caflisch, Lucius, "Regulation of the Uses of International Waterways: The Contribution of the United Nations" in Martin Ira Glassner (ed.), *The United Nations at Work* (Greenwood Publishing Group, 1998) 3–35.

Campbell, I.C., "Integrated Water Resources Management for the Mekong River Basin" in A.K. Biswas et al. (eds.) *Integrated Water Resources Management in South and South-East Asia* (Oxford University Press, 2005).

Caponera, Dante, "Patterns of Cooperation in International Water Law: Principles and Institutions" (1985) 25 *Nat. Resources J.* 563.

Carrard, Naomi, "Mainstream or Marginal? Transboundary Tributaries and the Mekong Agreement" (Presentation delivered at the 8th International Riversymposium, Brisbane, September 2005).

Chaipipat, Kulachada, *China Still Holds Trump Card on Use of Mekong Resources* (14 February 1993) The Nation.

Chaipipat, Kulachada, *Strong Distrust Delays Cooperation on Mekong* (27 March 1992) The Nation.

Chaipipat, Kulachada, *Mekong River Pact to Be Signed Today* (19 April 2000) The Nation.

Chambers, S., "Deliberative Democratic Theory" (2003) 6 *Annual Review of Political Science* 307.

Charoenpo, Ancha, *Megaproject is "a top priority" for new govt: Big plans for diversion of water from the Mekong* (4 February 2008) Bangkok Post.

Chea, V., "Water Supply in Phnom Penh" in *UN* Economic Commission for Asia and the Pacific, *Towards Efficient Water Use in Urban Areas in Asia and the Pacific*, (United Nations, 1998) ST/ESCAP/1874.

Chen, Huiping, Rieu-Clarke, Alistair, and Wouters, Patricia, "Exploring China's transboundary water treaty practice through the prism of the UN Watercourses Convention" (2013) 38 *Water Int'l* 217.

Choubert, Georges and Faure-Muret, Anne, *Geological Atlas of the World* (UNESCO, 1976).

Chu, T.H. et al., "Water, Climate, Food and Environment in the Mekong Basin in Southeast Asia" (Final Report, International Water Management Institute, Mekong River Commission and Institute of Environmental Studies, 2003).

Clark, M. et al., "Surface Uplift, Tectonics, and Erosion of Eastern Tibet from Large-Scale Drainage Patterns" (2004) 23 *Tectonics* TC1006, doi:10.1029/2002TC001402.

Cowie, Gail (ed.), "Reservoirs in Georgia: Meeting Water Supply Needs While Minimizing Impacts" (University of Georgia River Basin Science and Policy Center, May 2002).

CWPF-Mekong, *Dams of Cambodia in the Mekong Basin* (2013) <http://wle-mekong.cgiar.org/download/map-repository/A4_Cambodia_Element.pdf> (accessed 16 June 2017).

Crane, Mary, "It Becomes a River: Afghanistan at a Crossroads" (2009) 26(1) *World Policy Journal* 95.

Declaration on the Right to Development, A/RES/41/128, adopted by the General Assembly (4 December 1986).

Deener, David R. *The United States Attorney Generals and International Law* (The Hague, Martinus Nijhoff, 1957) 416.

De Stefano, Lucia et al., "Updating the International Water Events Database" (UNESCO Potential Conflict into Cooperation Potential (PCCP) Project, Dialogue Paper 7, 2009).

De Stefano, Lucia et al., "Tracking cooperation and conflict in international basins: historic and recent trends" (2010) 12(6) *Water Policy* 871.

Dinakaran, S. and Anbalagan, S., "Anthropogenic impacts on aquatic insects in six streams of south Western Ghats" (2007) 7 *Journal of Insect Science* 1.

Dinar, Ariel et al., *Bridges Over Water: Understanding Transboundary Water Conflict, Negotiation and Cooperation, World Scientific Series on Energy and Resource Economics* (World Scientific Publishing Co., 2007) vol. 3.

Dore, John, "The Governance of Increasing Mekong Regionalism" in M. Kaosa-ard and J. Dore (eds.), *Social Challenges for the Mekong Region* (White Lotus, 2003) 425.

Dore, John and Lazarus, Kate, "De-marginalizing the Mekong River Commission" in Francois Molle et al. (eds.), *Contested Waterscapes in the Mekong Region: Hydropower, Livelihoods and Governance* (Earthscan, 2009) 357.

Draft articles on the *Law of the Non-Navigational Uses of International Watercourses* and commentaries, Yearbook of the International Law Commission, vol. 2, part 2, at 91, art. 2(b), commentary 8 (1994).

Draft articles on the *Law of the Non-Navigational Uses of International Watercourses*, Yearbook of the International Law Commission, vol. 2, pt. 2, at 90 (1997).

Draft Articles on the *Responsibility of States for Internationally Wrongful Acts*, Report of the ILC on the Work of its Fifty-third Session, UN GAOR, 56th Sess., Supp. No. 10, p 43, UN Doc. A/56/10 (2001).

Duong, Wendy N., "Following the Path of Oil: The Law of the Sea or Realpolitik—What Good Does Law Do in the South China Sea Territorial Conflicts" (2007) 30 *Fordham Int. L. J.* 1098.

Economic Commission for Asia and the Far East (ECAFE), "Preliminary Report on Technical Problems Relating to Flood Control and Water Resources Development of the Mekong—an International River" (US Bureau of Reclamation, 1952).

Economic and Social Commission for Asia and the Pacific (ESCAP), "Atlas of Mineral Resources of the ESCAP Region" (Explanatory Brochure, United Nations, 1990) 19.

Economic and Social Commission for Asia and the Pacific (ESCAP), "Sources and Nature of Water Quality Problems in Asia and the Pacific," ST/ESCAP/1874 (United Nations, 1998).

Edelstein, David *Mr. Fiennes, I Presume?* (30 August 2005), Slate, <http://www.slate.com/id/2125293/?nav=fo> (accessed 23 June 2017).

Elhance, Arun P., *Hydro-Politics in the 3rd World: Conflict and Cooperation in International River Basins* (United States Institute of Peace Press, 1999).

Eliot, T.S., "Introduction," *reprinted in* Mark Twain, *The Adventures of Huckleberry Finn* (Norton Critical Edition, 1950).

Elleman, B. "Sino-Soviet Relations and the February 1979 Sino-Vietnamese Conflict" (Draft Paper, 20 April 1996) *in* 1996 Vietnam Symposium, *After the Cold War: Reassessing Vietnam* (Texas Tech University, Lubbock, Texas, USA, 18–20 April 1996).

Fawthrop, Tom, *Mekong Summit Struggles to Halt Devastating Dams* (9 April 2014) The Diplomat <http://thediplomat.com/regions/east-asia>.

Fawthrop, Tom, *Death by strangulation? Hydropower threatens to kill the mighty Mekong* (27 March 2015) The Ecologist http://www.theecologist.org/News/news_analysis/2806736/death_by_strangulation_hydropower_threatens_to_kill_the_mighty_mekong.html (accessed 19 September 2017).

Fernandes, Cristina Cox et al., "Amazonian Ecology: Tributaries Enhance the Diversity of Electric Fishes" (2004) 305 *Science* 1960.

Fravel, M.T., "Regime Insecurity and International Cooperation—Exploring China's Compromises in Territorial Disputes" (2005) 30*International Security* 46.

Freemen, Joshua D., "Taming the Mekong: The Possibilities and Pitfalls of a Mekong Basin Joint Energy Development Agreement" (2009) 10 *Asian-Pacific L. and Pol'y J.* 453.

Gao, Qi, *A Procedural Framework for Transboundary Water Management in the Mekong River Basin* (Brill Nijhoff, 2014) 37.

Gascon, Claude and Smith, Michael L., "Where Rivers Meet" (2004) 305 *Science* 1922.

Global Water Partnership, "Integrated Water Resources Management" (TAC Paper 4, Global Water Partnership Technical Committee, March 2000).

Gooch, Josefin, *The Mekong Agreement—In Light of International Principles Relating to Sustainable Use of Transboundary Watercourses* (Masters Thesis, Faculty of Law, University of Lund, Sweden, 2005).

Gottesman, Evan, *Cambodia After the Khmer Rouge: Inside the Politics of Nation Building* (Yale University Press, 2003).

Gupta, Avijit, "The Mekong River: Morphology, Evolution, Management" in Avijit Gupta (ed.), *Large Rivers: Geomorphology and Management* (John Wiley and Sons Ltd., 2007).

Gupta, Avijit, "Geology and Landforms of the Mekong Basin" in Ian C. Campbell (ed.), *The Mekong: Biophysical Environment of an International River Basin* (Academic Press, 2009) 29–51.

Gupta, Avijit and Liew, S.C., "The Mekong from Satellite Imagery: A Quick Look at a Large River," (2007) 85 *Geomorphology* 259.

Handley, Paul and Hiebert, Murray, *Hostile Undercurrents: Dispute Deepens Over Use of Mekong River Water* (2 April 1992) Far East Economic Review.

Harbinson, Rod, *Cambodian Dam Proceeds Despite Opposition over Fish, Ousted Villagers* (17 August 2015) Mongabay <https://news.mongabay.com/2015/08/cambodian-dam-proceeds-despite-opposition-over-effects-on-fish-ousted-villagers/> (accessed 20 June 2017).

Hawkesworth, N. et al., "Independent Organisational Financial and Institutional Review of the MRC Secretariat and the National Mekong Committees" (MRC, Vientiane, Lao PDR, 2007).

Hearings on H.R. 5773 before the House Committee on Irrigation and Reclamation (United States) 70th Cong., 1st Sess., 1927.

Hensengerth, Oliver, "Transboundary River Cooperation and the Regional Public Good: The Case of the Mekong River" (2009) 31(2) *Contemporary Southeast Asia* 326.

Hirsch, Philip, "Beyond the Nation State: Natural Resource Conflict and 'National Interest' in Mekong Hydropower Development" (1999) 29 *Golden Gate Univ. L. Rev.* 399.

Hirsch, Philip, "The Mekong River Commission and the Question of National Interest(s)" (2006) 12(1) *Watershed* 20.

Hirsch, Philip, "The Changing Political Dynamics of Dam Building on the Mekong" (2010) 3(2) *Water Alternatives* 312.

Hirsch, Philip and Cheong, Gerard, "Natural Resource Management in the Mekong River Basin: Perspectives for Australian Development Cooperation" (Final Overview Report to AusAID, University of Sydney, 1996).

Hirsch, Philip and Wyatt, Andrew, "Negotiating Local Livelihoods: Scales of Conflict in the Se San River Basin" (2004) 45 *Asia Pacific Viewpoint* 51.

Hirsch, Philip and Jensen, Kurt Morck *with* Ben Boer, Naomi Carrard, Stephen FitzGerald and Rosemary Lyster, "National Interests and Transboundary Water Governance in the Mekong" (Australian Mekong Resource Centre (AMRC), University of Sydney in collaboration with Danish International Development Assistance, May 2006).

Hoey, T. et al., "The Importance of Tributary Sediment Supply for the Sediment Dynamics of a Large Regulated Gravel-Bed River" (abstract, 2004).

Hori, Hiroshi, *The Mekong: Environment and Development* (United Nations University Press, 2000).

Hutchinson, C.S., *Geological Evolution of Southeast Asia*, Oxford Monographs on Geology and Geophysics no. 13. (Clarendon Press, 1989) 368.

Interim Mekong Committee, "Preparing for New Challenges" (1991 Interim Mekong Committee Annual Report, Bangkok, Thailand).

The Dublin Statement on Water and Sustainable Development (Dublin Principles) (International Conference on Water and the Environment, Dublin, Ireland, 26–31 January 1992).

International Joint Commission, "Impacts of a Proposed Coal Mine in the Flathead River Basin" (1988).

International Rivers Network, *Damming the Sesan River: Impacts on Cambodia and Vietnam* (October 2002) <https://www.internationalrivers.org/resources/damming-the-sesan-river-impacts-in-cambodia-and-vietnam-2673> (accessed 19 September 2017).

International Rivers Network, "Sizing up the Grid: How the Mekong Power Grid Compares Against the Policies of the Asian Development Bank" (2004).

International Rivers Network, *Mekong Mainstream Dams* (2016), available at: <http://internationalrivers.org/en/southeast-asia/mekong-mainstream-dams> (accessed 13 June 2017).

International Rivers Network, *Sesan River*, available at <http://www.internationalrivers.org/en/node/1039> (accessed 21 June 2017).

International Rivers Network and American Rivers, "Beyond Dams: Options and Alternatives" (2004).

International Union for Conservation of Nature (IUCN), "A Window of Opportunity for the Mekong Basin: The UN Watercourses Convention as a Basis for Cooperation (A Legal Analysis of How the UN Watercourses Convention Complements the Mekong Agreement)" (IUCN, 2016).

International Water Management Institute, "Water Governance in the Mekong Region: The Need for More Informed Policy-Making" (Water Policy Briefing Issue No. 22, November 2006).

Jacobs, Jeffrey W., "Towards Sustainability in Lower Mekong River Basin Development" (1994) 19 *Water International* 43.

Johannesburg Plan of Implementation (2002), available at <http://www.un.org/esa/sustdev/documents/WSSD_POI_PD/English/WSSD_PlanImpl.pdf> (accessed 29 June 2017).

Johnson, Thomas H. and Mason, M. Chris, *All Counterinsurgency Is Local* (October 2008) The Atlantic 36.

Kamkongsak, L. and Law, M., "Laying Waste to the Land: Thailand's Khon-Chi-Mun Irrigation Project" (2001) 6(3) *Watershed* 25.

Kanwerayothin, Supapohn, *The Mekong: More of a Liability Than An Asset?* (21 March 1992) Bangkok Post.

Kaya, Ibrahim, "The Euphrates-Tigris Basin: An Overview and Opportunities for Cooperation under International Law," (Arid Lands Newsletter no. 44, Fall/Winter 1998), available at <http://ag.arizona.edu/OALS/ALN/aln44/kaya.html> (accessed 15 June 2017).

Kensler, Mike, Director of Outreach Programs, Auburn University Water Center (email to author, 11 May 2010, 9:24 a.m. CST).

Khalid, Abu Raihan M., "The Interlinking of Rivers Project in India and International Water Law: An Overview" (2004) 3 *Chinese J. Intl. L.* 553.

Khanna, P., *The Second World: Empires and Influence in the New Global Order* (Random House, 2008).

Khruschev Report on Moscow Conference of Representatives of Communist and Workers Parties (Folder 4, President's Office Files Box 126(a), John F. Kennedy Library, Boston, MA, 6 January 1961).

Kinna, Remy and Rieu-Clarke, Alistair, "The Governance Regime of the Mekong River Basin" (2017) 2 *Int'l Water L.* 1.

Kiss, Alexandre and Shelton, Dinah, *International Environmental Law* (Transnational Publishers, Inc., 2004).

Kitamura, Masayuki, "Discovery and First Descent of the Mekong Headwaters" (October 2001) 1 *Japanese Alpine News* 1.

Krasner, Stephen D., "Structural Causes and Regime Consequences: Regimes as Intervening Variables" in Stephen D. Krasner (ed.), *International Regimes* (Cornell University Press, 1986) 2.

Krutilla, John V., *The Columbia River Treaty: The Economics of an International River Basin Development* (John Hopkins Press, 1967).

Lammers, J., *Pollution of International Watercourses* (Matinus Nijhoff Publishers, 1984).

Lan, Mai, *Diverting the Mekong River into Thailand: The Khong-Loei-Chi-Mun Project* (6 June 2016) Mekong Commons <http://www.mekongcommons.org/large-scale-irrigation-northeastern-thailand-reprised-khong-loei-chi-mun-project/> (accessed 20 June 2017).

Lang, Chris, "The Legacy of Savage Development: Colinisation of Vietnam's Central Highlands" (November 1995–February 1996) 1(2) *Watershed* 43.

Law on Water and Water Resources (Lao People's Democratic Republic) National Assembly, 11 October 1996, art. 25.

Law on Water Resources Management (Kingdom of Cambodia), National Assembly, 29 June 2007, arts. 1, 4, 7, 8, 19.

Lazarus, Kate (ed.) et al., "Lancang-Mekong: A River of Controversy" (IRN and Mekong Watch, 2003).

Lebel, Louis and Garden, Po, "Deliberation, Negotiation and Scale in the Governance of Water Resources in the Mekong Region" in C. Pahl-Wostl et al. (eds.), *Adaptive and Integrated Water Management* (Springer, 2008) 216.

Lebel, Louis et al., "The Politics of Scale, Position, and Place in the Governance of Water Resources in the Mekong Region" (2005) 10(2) *Ecology and Society*, article 18, available at <http://www.ecologyandsociety.org/vol10/iss2/art18/> (accessed 12 June 2017).

Lebel, Louis et al., "Beyond 'Basin': The Politics of Scale in Mekong Water Governance" (CGIAR Science Paper PA079, 2008), available at <https://www.researchgate.net/profile/Theib_Oweis/publication/266891806_A_methodology_for_the_assessment_of_agricultural_water_productivity_at_the_river_basin_level/links/543e4b670cf240f04d10f152.pdf#page=71> (accessed on 15 June 2017).

Lerner, Michael, "Draft Se San Legal and Policy Study" (Oxfam America, 2002a).

Lerner, Michael, "Dangerous Waters: Violations of International Law and Hydropower Development along the Sesan River" (Background Briefing 4, National Sesan River Protection Network Workshop, 27 November 2002b).

Lerner, Michael, "Dangerous Waters: Violations of International Law and Hydropower Development Along the Sesan River" (unpublished manuscript, Oxfam, Phnom Penh, Cambodia, December 2003).

Lien, N.D., "Catchment Areas and Average Yearly Flows of Major Tributaries at Their Confluence with the Mekong" (May 1983) reprinted in V.R. Pantulu, "The Mekong River System" in Bryan R. Davies and Keith F. Walker (eds.), *The Ecology of River Systems* (Springer, 1986) 708.

Lipper, Jerome, "Equitable Utilization" in A. Garretson et al. (eds.), *The Law of International Drainage Basins* (Oceana Publications, 1968).

Liu, H. "Water Resources Simulation Model of the Lancang River (Yunnan Portion)" in D. He, G. Zhang, and K. Hsiang-te (eds.), *Towards Cooperative Utilization and Coordinated Management of International Rivers* (Science Press, 2001) 208–215.

Lorenzetti, E.A. et al., "Structural Styles in Rift Basins: Interpretation Methodology and Examples from Southeast Asia" (1994) 78 *AAPG Bulletin* 1152.

Louka, Elli, *Water Law and Policy* (Oxford University Press, 2008).

Loures, Flavia, World Wildlife Fund International Water Law and Policy Senior Program Officer (email to author, 4 April 2009, 11:04 a.m.CST) (on file with author).

Loures, Flavia and Dellapenna, Joseph, "Forthcoming Developments in International Groundwater Law: Proposals for the Way Ahead" (August 2007) 21 *Water Environment* 58–62.

Lowe-McConnell, R., *Ecological Studies in Tropical Fish Communities* (Cambridge University Press, 1987) 382.

Lower Mekong Initiative, *The Lower Mekong Initiative (LMI)* <http://lowermekong.org/about/lower-mekong-initiative-lmi> (accessed 23 June 2017).

Makim, Abigail, "The Changing Face of Mekong Resource Politics in the Post-Cold War Era: Re-negotiating Arrangements for Water Resource Management in the Lower Mekong River Basin (1991–1995)" (Working Paper No. 6, Australian Mekong Resource Centre, University of Sydney, August 2002).

Manager Magazine, "Thai and Vietnam Disagree to Investigate the Lower Mekong Basin" (translated from the Thai periodical in Browder (2000), 28 March 1994).

McCaffrey, Stephen C. (Special Rapporteur), *Second Report on the Law of the Non-Navigational Uses of International Watercourses*, U.N. Doc. A/CN.4/399 (1986) Par. 170.

McCaffrey, Stephen C., "International Organizations and the Holistic Approach to Water Problems" (1991) 31 *Nat. Resources J.* 139.

McCaffrey, Stephen C., "The Harmon Doctrine One Hundred Years Later: Buried, Not Praised" (1996) 36 *Nat. Resources J.* 549.

McCaffrey, Stephen C., "An Overview of the U.N. Convention on the Law of the Non-Navigational Uses of International Watercourses" (2000) 20 *J. Land Resources and Envtl L.* 57.

McCaffrey, Stephen, "International Water Law for the 21st Century: The Contribution of the U.N. Convention" (January 2001) 118 *Water Resources Update* 11.

McCaffrey, Stephen C., "The Contribution of the UN Convention on the Law of the Non-Navigational Uses of International Watercourses" (2001) 1 *Int. J. Global Envtl. Issues* 250.

McCaffrey, Stephen C., *The Law of International Watercourses*, (Oxford University Press, 2nd ed., 2007).

McCaffrey, Stephen C., "Introduction, Symposium: Taking Stock of Sustainable Development at 20: A Principle at Odds with Itself?" (2008) 21 *Pac. McGeorge Global Bus. and Dev. L.J.* 151.

McCaffrey, Stephen, "Establishing Rights—The Law of International Waters" in Aaron T. Wolf (ed.), *Sharing Water, Sharing Benefits: Working Towards Effective Transboundary Water Resources Management* (A Graduate/Professional Skills-Building Workbook, adapted from the World Bank International Waters Course, 2010a) 192–193.

McCaffrey, Stephen C., "Sovereign and Cooperative Management of Shared Water Resources in a Time of Shrinking Availability: The Role of International Law, Transboundary Aquifers: Challenges and New Directions" (Abstracts, UN Educational, Scientific and Cultural Organization (UNESCO) International International Hydrological Programme (IHP), ISARM 2010 International Conference, Paris, 6–8 December 2010b) 114.

McCaffrey, Stephen C. and Sinjelz, Mpazi, "The 1997 United Nations Convention on International Watercourses" (1998) 92 *Am. J. Intl. L.* 97.

McCaffrey, Stephen C. and Weber, Gregory S., *Guidebook for Policy and Legislative Development on Conservation and Sustainable Use of Freshwater Resources* (United Nations Environment Programme (UNEP), Nairobi, Kenya, 2005).

McIntyre, Owen, *Environmental Protection of International Watercourses Under International Law* (Ashgate Publishing Ltd., 2007).

Meade, R.H. et al., "Backwater Effects in the Amazon River Basin in Brazil" (1991) 18 *Environmental Geology and Water Science* 105.

Mekong Committee, *Report on Indicative Basin Plan: A Proposed Framework for the Development of Water and Related Resources of the Lower Mekong Basin*, E/CN.11/WRD/MKG/L.340, Bangkok, Thailand (UN, 1970).

Mekong Committee, "Draft Statement of Principles for the Planning and Utilization of the Waters of the Lower Mekong Basin" WRD/MKG/INF/L.466, Bangkok, Thailand (1971).

Mekong Committee, "The 1973 Mekong Committee Report" (1973).

Mekong Committee, "The Mekong Committee: A Historical Account (1957–1989)" (Mekong Committee Secretariat, Bangkok, Thailand, 1989).

Mekong River Commission, *Lower Mekong Hydrologic Yearbooks* (MRC, 1994–2004).

Mekong River Commission, "Mekong River Commission: Towards Sustainable Development" (Annual Report, 1995).

Mekong River Commission, *Lower Mekong Hydrologic Yearbook* (MRC, 1998a) 19–23.

Mekong River Commission, "Water Utilization Program Preparation Project: Final Report," MKG/R.98/013 (1998b).

Mekong River Commission, Mekong River Basin diagnostic study—Final report, Mekong River Commission, Bangkok Thailand, MKG/R 97010 (1997).

Mekong River Commission, Fax from the Mekong River Commission and accompanying "Report on visit to Ialy[Yali] Hydropower Project 20–21 April 2000" (15 May 2000).

Mekong River Commission, Report Regarding Terms of Reference for Studies on Se San River Water Management (2001a).

Mekong River Commission, 1995 Mekong Agreement and Procedural Rules, Procedures for Data and Information Exchange and Sharing Bangkok, Thailand, at 20–24 (1 November 2001b).

Mekong River Commission, Deep pools as dry season fish habitats in the Mekong River Basin, Technical Paper No. 4, Report prepared by the MRC Fisheries Programme at the request of the Technical Advisory Body on Fisheries Management in the Lower Mekong Basin, Phnom Penh, Cambodia, 22 pp. (9 April 2002a).

Mekong River Commission, "Mekong News" (Newsletter of the MRC, April–June 2002b (issue no. 2)).

Mekong River Commission, 1995 Mekong Agreement and Procedural Rules, Procedures for Notification, Prior Consultation and Agreement (PNPCA), Phnom Penh, Cambodia, at 38–54 (30 November 2003a).

Mekong River Commission, 1995 Mekong Agreement and Procedural Rules, Procedures for Water Use Monitoring, Phnom Penh, Cambodia, at 31–36 (30 November 2003b).

Mekong River Commission 2004 Annual Report, Vientiane, Lao PDR (2005a).

Mekong River Commission, Overview of the Hydrology of the Mekong Basin, Vientiane, Lao PDR (November 2005b).

Mekong River Commission, Procedures for Maintenance of Flows on the Mainstream, Ho Chi Minh City, Vietnam (22 June 2006).

Mekong River Commission, Procedures for Water Quality (endorsed by MRC Joint Committee in 2006, pending approval by MRC Council).

Mekong River Commission, "Strategic Plan: 2006–2010" (December 2006).

Mekong River Commission, Comments by Mr. Do Manh Hung, Officer-in-Charge, MRC Secretariat (Press Statement regarding the press briefing held on 13 November in Bangkok, Thailand by TERRA, 15 November 2007).

Mekong River Commission, Letter from MRC Secretariat CEO Jeremy Bird to Khun Premrudee, Requests from Mekong Civil Society (11 April 2008a), available at: <http://www.MRCmekong.org/assets/Publications/Events/MRCS-CEO-Responses-TERRA.pdf> (accessed 30 June 2017).

Mekong River Commission, "High Level Study Visit to the Columbia River Basin: Back-to-Office Report" (MRC Basin Development Plan Programme, Phase 2, June 2008b).

Mekong River Commission, "Mekong News" (Newsletter of the Mekong River Commission, July–October 2008c (issue no. 3)).

Mekong River Commission, *Hydropower Database*, (MRC, 2010a).

Mekong River Commission, "State of the Basin Report 2010" (MRC, April 2010b).

Mekong River Commission, *Public Participation in the Context of the MRC*, <http://www.MRCmekong.org/assets/Publications/policies/Public-Participation-in-MRC-context.pdf> (accessed 21 June 2017).

Mekong River Commission, *The Story of Mekong Cooperation* <http://www.MRCmekong.org/about-MRC/history/> (accessed 12 June 2017).

Mekong River Commission, *The Mekong Programme*, "Table 1.1: Approximate Distribution of MRB Water Resources by Country" <http://archive.iwlearn.net/www.MRCmekong.org/www.MRCmekong.org/mekong_program_ceo.htm> (accessed 8 June 2017).

Mekong River Commission, *MRC appoints new Chief Executive Officer* (Press Release, 14 November 2011) <http://www.MRCmekong.org/news-and-events/news/MRC-appoints-new-chief-executive-officer/> (accessed 19 September 2017).

Mekong River Commission, *Lancang-Mekong Cooperation: MRC Welcomes the New Initiative for Regional Cooperation by Six Countries in the Mekong River Basin* (Press Release, 31 March 2016a) <http://www.MRCmekong.org/news-and-events/news/lancang-mekong-cooperation-MRC-welcomes-the-new-initiative-for-regional-cooperation-by-six-countries-in-the-mekong-river-basin/> (accessed 19 September 2017).

Mekong River Commission, "Strategic Plan 2016–2020" (March 2016b) 9.

Mekong River Commission, *First Riparian Chief Executive Officer Assumes His Office Today* (Press Release, 18 January 2016c) <http://www.MRCmekong.org/news-and-events/news/first-riparian-chief-executive-officer-assumes-his-office/> (accessed 18 September 2017).

Mekong River Commission Council, "Integrated Water Resources Management-based Basin Development Strategy 2016–2020 For the Lower Mekong Basin" (2016).

The Mekong Reconnaissance Team, "Comprehensive Reconnaissance Report on the Major Tributaries of the Lower Mekong Basin" (Mekong Committee, 1961).

Mekong Sources, "Mekong—The Mother of Waters" (2003).

"Memorandum in Relation to the Arbitration of the *Trail Smelter* Case, United States and Canada," 10 August 1937 (prepared by Green H. Hackworth, Legal Advisor, for Swagar Sherley, Agent of the United States) in Whiteman, *Digest of International Law*, (US Department of State, vol. 5, 1963–1970) 60–61.

"Memorandum to the Legal Advisor" (23 November 1942) reprinted in Marjorie M. Whiteman, *Digest of International Law* (US Department of State, vol. 3 1964) 950.

Menon, P.K., "Financing the Lower Mekong River Basin Development" (1972) 44 *Pacific Affairs* 566.

Merriam-Webster Dictionary <http://www.merriam-webster.com/dictionary/basin> (accessed 14 June 2017).

Meyer, J.L. and Wallace, J.B., "Lost Linkages and Lotic Ecology: Rediscovering Small Streams" in M.C. Press et al. (eds.), *Ecology: Achievement and Challenge: 41st Symposium of the British Ecological Society* (jointly sponsored by the Ecological Society of America, held at Orlando, Florida, usA, 10–13 April 2000; Cambridge University Press, 2001).

Molle, Francois, "Irrigation and Water Policies in the Mekong Region: Current Discourses and Practices" (IWMI Research Report 95, 2005).

Molle, Francois and Floch, Philippe, "Water, Poverty and the Governance of Mega Projects: The Thai 'Water Grid'" (IWMI Working Paper, 2007).

Molle, Francois and Floch, Philippe, "Megaprojects and Social and Environmental Changes: The Case of the Thai 'Water Grid'" (2008a) 37 *Ambio* 199.

Molle, Francois and Floch, Philippe, "The 'Desert Bloom' Syndrome: Politics, Ideology, and Irrigation Development in the Northeast of Thailand" (IWMI Working Paper 4, 2008b).

Molle, Francois et al., "Contested Mekong Waterscapes: Where to Next?" in Francois Molle et al. (eds.), *Contested Waterscapes in the Mekong Region: Hydropower, Livelihoods and Governance* (Earthscan, 2009a) 383–413.

Molle, Francois et al., "Introduction: Changing Waterscapes in the Mekong Region-Historical Background and Context" in Francois Molle et al. (eds.), *Contested Waterscapes in the Mekong Region: Hydropower, Livelihoods and Governance* (Earthscan, 2009b) 1–19.

Mouth to Source, *3SPN-Dams on Sesan River Polluting Water and Poisoning Downstream Communities* (8 September 2009) <http://khmerization.blogspot.com/2009/09/new-scientific-study-dams-on-sesan.html> (accessed 16 June 2017).

Moyle, Peter B. and Mount, Jeffrey F., "Homogenous Rivers, Homogenous Faunas" (3 April 2007) 104 *Proceedings of the National Academies of Sciences of the U.S.A.* (PNAS) 5711.

Muckleston, Keith W., "International Management in the Columbia River System" (Technical Document in Hydrology, UNESCO SC-2003/WS/56, 2003).

Nakayama, Mikiyasu, "Aspects Behind Differences in Two Agreements Adopted by Riparian Countries of the Lower Mekong River Basin" (1999) 1 *J. Comp. Policy Anal.* 293.

The Nation, *Infrastructure Project: Tap water grid planned by '05* (23 June 2006).

National Research Council of the National Academies, *Colorado River Basin Water Management: Evaluating and Adjusting to Hydroclimatic Variability* (National Academies Press, 2007).

NEDECO, "Mekong Delta Master Plan: Thematic Study on Management of Water Resources," "Volume 1: Optimal Use of Water Resources" (MRC, Bangkok, Thailand, October 1993).

NGO Forum on Cambodia, "Down River: The Consequences of Vietnam's Se San River Dams on Life in Cambodia and Their Meaning in International Law" (prepared by Eric Rutkow, Cori Crider, and Tyler Giannini, December 2005) 8. *Available at* http://hrp.law.harvard.edu/wp-content/uploads/2013/02/Down-River-2005.pdf (accessed 19 September 2017).

Nguyen, Quang M., "Hydrologic Impacts of China's Mekong Dams on Lower Mekong River" (2003).

Noe, Susan Y., "Belize's Chalillo Dam: A Sustainability Analysis of a Proposed Development Project" (unpublished manuscript, 4 April 2002).

Oregon State University Institute for Water and Watersheds, *International Water Events Database: 1950–2008*, Available at <http://www.transboundarywaters.orst.edu/database/interwatereventdata.html> (accessed 23 June 2017).

Osborn, Rachael Paschal, "Columbia River Treaty: Past and Possible Futures" (WA-AWRA Section Newsletter, November–December 2007) 9.

Osborne, Milton, "River at Risk: The Mekong and the Water Politics of China and Southeast Asia" (Lowy Institute for International Policy Paper 02, Longueville Media, New South Wales, Australia, 2004).

Osborne, Milton, *Under-Threat Mekong River Faces New Dam Challenges* (12 November 2008) The Nation.

Paisley, Richard, "Adversaries Into Partners: International Water Law and the Equitable Sharing of Downstream Benefits" (2002) 3(2) *Melbourne J. Int. Law* 280.

Pantulu, V.R., "The Mekong River System" in Bryan R. Davies and Keith F. Walker (eds.), *The Ecology of River Systems* (Springer, 1986a) 706.

Pantulu, V.R., "Fish of the Lower Mekong Basin" in Bryan R. Davies and Keith F. Walker (eds.), *The Ecology of River Systems* (Springer, 1986b) 721.

Pearce, Fred, *When the Rivers Run Dry: Water—The Defining Crisis of the Twenty-First Century* (Beacon Press, 2006).

Petkova, E., and Veit, P., "Environmental Accountability Beyond the Nation-State: The Implications of the Aarhus Convention" (Environmental Governance Note, World Resources Institute, 2000).

Peissel, Michel, *The Last Barbarians: The Discovery of the Source of the Mekong in Tibet* (Henry Holt and Company, 1997).

Peltzer, G. and Tapponier, P., "Formation and Evolution of Strike-Slip Faults, Rifts and Basins During the India-Asia Collision: An Experimental Approach" (1988) 93 *Journal of Geophysical Research* 15085.

Peterson, B.J. et al., "Stream Processes Alter the Amount and Form of Nitrogen Exported from Small Watersheds" (2001) 292 *Science* 86.

Phillips, David et al., "Trans-boundary Water Co-operation as a Tool for Conflict Prevention and Broader Benefit Sharing" (prepared for the Sweden Ministry for Foreign Affairs, 2006).

Pichyakorn, Bantita, "Sustainable Development and International Watercourse Agreements: The Mekong and the Rhine" (International Union for Conservation of Nature (IUCN) Draft Paper, 30 June 2002).

Pichyakorn, Bantita, "International Watercourses Law: The Experience of the Mekong River Basin" in Surya P. Subedi (ed.), *International Watercourses Law for the 21st Century: The Case of the River Ganges Basin* (Ashgate Publishing, 2005).

Poff, N.L. et al., "The Natural Flow Regime" (1997) 47 *Bioscience* 769.

Putt, Larry O., "Water Resources Protection in Alabama: The Need for a Paradigm Change" (2003) 7 *T.G. Jones L. Rev.* 1.

Puustjarvi, Esa, "Review of Policies and Institutions Related to Management of Upper Watershed Catchments" (Report of Greater Mekong Subregion Watershed Project (Phase II): Poverty Reduction and Environmental Management in Remote Greater Mekong Sub-region, 2004).

Radosevich, George, "Third Meeting of the Mekong Working Group (MWG-III): Report on Country Position Papers, Findings, Conclusion, Proposed Draft Agreement, and Meeting Results" (United Nations Development Programme, Bangkok, Thailand, 7 July 1993a) [Radosevich (1993a)].

Radosevich, George, "Draft Agreement From Technical Drafting Meeting (TDM)" (UNDP, Bangkok, Thailand, 4 September 1993b) [Radosevich (1993b)].

Radosevich, George, "MWG-IV Report and Request for Action" (UNDP, Bangkok, Thailand, 12 December 1993c) 1 [Radosevich (1993c)].

Radosevich, George, "Minutes of the TDM-II" (UNDP, Bangkok, Thailand, 6–7 January 1994).

Radosevich, George E., "Agreement on the Cooperation for the Sustainable Development of the Mekong River Basin: Commentary and History" (UNDP, 1995).

Radosevich, George E. and Olson, Douglas C., "Existing and Emerging Basin Arrangements in Asia: Mekong River Commission Case Study" (Third Workshop on River Basin Institution Development, The World Bank, Washington, DC, 24 June 1999).

Räsänen, Timo A. et al., "Observed River Discharge Changes Due to Hydropower Operations in the Upper Mekong Basin" (2017) 545 *J. of Hydrology* 28, <https://www.thethirdpole.net/2017/02/06/new-study-shows-significant-impact-of-chinese-dams-on-mekong/> (accessed 13 June 2017).

Revenga, Carmen et al., *Watersheds of the World: Ecological Value and Vulnerability* (World Resources Institute, 1998).

"Report of the Commission to the General Assembly on the work of its thirty-second session" (1980) *Yearbook of the International Law Commission*, vol. 2, part 2, at 110–111.

Rice, Stephen, Department of Geography, Loughborough University, England, UK (email communiqué to author, 19 May 2009, 2:18 a.m. PDT).

Rice, S.P. et al., "Tributaries, Sediment Sources, and the Longitudinal Organization of Macroinvertebrate Fauna Along River Systems" (April 2001) 58 *Canadian J. of Fisheries and Aquatic Sciences* 824.

Rice, Stephen et al. (eds.), *River Confluences, Tributaries and the Fluvial Network* (Wiley, 2008).

Rieu-Clarke, Alistair, "Entry into Force of the 1997 UN Watercourses Convention: Barriers, Benefits and Prospects" (December 2007) *Water 21* 12.

Rieu-Clarke, Alistair, "Notification and Consultation Procedures Under the Mekong Agreement: Insights from the Xayaburi Controversy" (2015) 5(1) *Asian J. of Int'l L.* 143.

Rieu-Clarke, Alistair and Gooch, Geoffrey, "Governing the Tributaries of the Mekong—The Contribution of International Law and Institutions to Enhancing Equitable Cooperation Over the Sesan" (2010) 22(2) *Pac. McGeorge Global Bus. and Dev. L. J.* 193.

Ringler, Claudia, *Optimal Allocation and Use of Water Resources in the Mekong River Basin: Multi-Country and Intersectoral Analyses* (Development Economics and Policy Series, Peter Lang, 2001) vol. 20.

Rio Declaration, UN Doc. A/CONF.151/5/Rev. 1, 13 June 1992, 31 ILM 874 (UN Conference on Environment and Development (Earth Summit), Rio de Janeiro, Brazil, 3–14 June 1992).

Rivers Watch, "Laos-Thai Friendship Water Diversion Project" (2005).

Rix, Adam S., "The Mekong River Basin: A Resource at the Cross-Roads of Sustainable Development" (2003) 21 *Temp. Envtl. L. and Tech. J.* 103.

Roy, D., "Hegemon on the Horizon? China's Threat to East Asian Security" (1994) 19 *International Security* 149.

Rutkow, Eric et al., "Down River: The Consequences of Vietnam's Se San River Dams on Life in Cambodia and Their Meaning in International Law" (NGO Forum on Cambodia, Phnom Penh, Cambodia, December 2005).

Ryder, Grainne, *Ending Vietnam's Hydro Threat to Cambodia's Mekong Tributaries: Why Power Sector Reform Matters* (Probe International, November 2002).

Sands, Philippe, *Principles of International Environmental Law* (Cambridge University Press, 2nd ed., 2003).

Sangha, Kim and Bunnarith, Tep, "Lessons Learned But Not Learned: Water Governance in the 3S Rivers Region" (Vientiane, Lao PDR, 6–7 July 2006).

Sauser-Hall, Georges, *L'utilisation Industrielle des Fleuves Internationaux* (Brill, Recueil des Cours, Tome 83, No. 2, 1953) 477–481.

Schrier, Hans, University of British Columbia Institute for Resources, Environment and Sustainability (email communiqué to author, 6 November 2007, 11:27 a.m. CST).

Schwebel, Stephen M. (Special Rapporteur), *Third Report on the Law of the Non-Navigational Uses of International Watercourses,* A/CN.4/348 (1 December 1981).

Segal, G., "East Asia and the Constrainment of China" (1996) 20 *International Security* 107.

Se San Protection Network (SPN), "Learning from Transboundary Environmental Conflicts" (October–December 2003) 6(4) *Mekong Update and Dialogue* 5.

SPN, Ratanakiri Se San Communities, "Joint Statement from Ethnic Minority Groups Living Along the Sesan River in Ratanakiri" (Ratanakiri Province, Cambodia, 2002).

Shmueli, Deborah F., "Water Quality in International River Basins" (1999)18 *Political Geography* 437.

Sindorf, Nikolai and Dangthuy, Trang, "Aquatic Ecosystem Connectivity, Ecosystem Processes and Watershed Integrity under Mekong Dam Development" (MRC International Conference on Watershed Management, Chiang Mai, Thailand, 9–11 March 2011).

Sneddon, Chris and Fox, Coleen, "Rethinking Transboundary Waters: A Critical Hydropolitics of the Mekong Basin" (2006) 25 *Political Geography* 181.

Snidvongs, Anond and Teng, Seng-Keh, "Mekong River" (GIWA Regional Assessment 55, Global International Waters Assessment, 2006).

Southeast Asia Rivers Network (SARN), "A Summary Report: Conceptual Study Report on Laos-Thai Friendship Water Development for Sustainable Agriculture in Savannakhet Province of Lao P.D.R. and Lower Chi Basin of Thailand" (conducted by Sanyu Consultants Inc., June 1998; prepared by Southeast Asia Rivers Network-Thailand Chapter, 20 June 2002).

Sriganesh, Bhargav, *Compliance with UN Watercourses Convention: Half Full or Half Empty?* (unpublished dissertation, Kings College, London, 12 May 2017). *available at* http://www.e-ir.info/2017/05/12/compliance-with-un-watercourses-convention-half-full-or-half-empty/ (accessed 23 June 2017).

Declaration of the United Nations Conference on the Human Environment (Stockholm Declaration), Principle 21, 11 ILM 1416 (16 June 1972).

Suong, Thu, *Mekong Basin Stirs up Region: Thai Water Diversion Project Could have Mega Risks*, (8 July 2016) Mekong Eye <https://www.mekongeye.com/2016/07/08/mekong-basin-stirs-up-region-thai-water-diversion-project-could-have-mega-risks/> (accessed 20 June 2017).

SWECO Groner (in association with Norwegian Institute for Water Research, ENVIRO-DEV and ENS Consult), "Environmental Impact Assessment on the Cambodian part of the Se San River due to Hydropower Development in Vietnam Electricity of Viet Nam (EVN), Power Engineering Consulting Company No. 1" (December 2006).

Tandon, S.K. and Sinha, R., "Geology of Large River Systems" in A. Gupta (ed.), *Large Rivers: Geomorphology and Management* (John Wiley and Sons Ltd., 2007) 7–28.

Tapponier, P. et al., "Propagating Extrusion Tectonics in Asia, New Insights from Simple Experiments with Plasticine" (1982) 10 *Geology* 611.

Tapponier, P. et al., "On the Mechanics of the Collision between India and Asia" in M.P. Coward and A.C. Ries (eds.), *Collision Tectonics* (Geological Survey of London Special Publication 19, 1986) 115–157.

Tarlock, A. Dan, "International Water Allocation," sec. 11:14 in *Law of Water Rights and Resources* (Thomson Reuters, 2001).

Teclaff, Ludwik A., *The River Basin in History and Law* (The Hague/Martinus Nijhoff, 1967).

Tesli, A. and Husby, S.R., "EIA in a Transboundary Context: Principles and Challenges for a Coordinated Nordic Application of the Espoo Convention" (1999) 19 *Environmental Impact Assessment Review* 57.

Thim, Ly, *Planning the Lower Mekong Basin: Social Intervention on the Se San River* (Transaction Publishers, 2010) 125, 153.

Thorson, Erica J., "Sharing Himalayan Glacial Meltwater: The Role of Territorial Sovereignty" (2009) 19 *Duke J. Comp. and Intl. L.* 487.

Tingsanchali, Tawatchai and Singh, Purushottam Raj, "Optimum Water Resources Allocation for Mekong-Chi-Mun Transbasin Irrigation Project, Northeast Thailand" (1996) 21 *Water International* 20.

Tiodolf, Anna Madeleine, *A Limnological Study in the Se San River in Cambodia during the Dry Season: Focus on Toxic Cyanobacteria and Coliform Bacteria* (unpublished Master thesis, Norwegian University of Life Sciences, 2009).

Tiodolf, Anna Madeleine and Stålnacke, Per, "A Limnological Study in the Se San River in Cambodia During the Dry Season: Focus on Toxic Cyanobacteria and Coliform Bacteria" (STRIVER Technical Brief No. 12, 2009).

Titthara, May, *Trapped between Two Dams* (7 January 2016) Khmer Times <http://www.khmertimeskh.com/news/19609/trapped-between-two-dams/> (accessed 18 June 2017).

Toda, Osamu et al., "Evaluation of Tributaries Contribution in the Mekong River Basin During Rainy and Dry Seasons" (Paper presented at the second conference of the

Asia Pacific Association of Hydrology and Water Resources, Singapore 5–9 July 2004).

Trandem, Ame, "A Vietnamese/Cambodian Transboundary Dialogue: Impacts of dams on the Se San River" (2008) 51 *Development* 108.

Twidale, C., "River Patterns and Their Meanings" (2004) 67 (3–4) *Earth Science Reviews* 159.

United Nations, *Convention on the Law of the Non-Navigational Uses of International Watercourses*, https://treaties.un.org/pages/viewdetails.aspx?src=ind&mtdsg_no=xxvii-12&chapter=27&lang=en (status as of 28 June 2017).

UNEP, *Hydropolitical Vulnerability and Resilience along International Waters: Asia* (UNEP, 2009).

UN General Assembly resolution 1401, *Preliminary Studies on the Legal Problems Relating to the Utilization and Use of International Rivers*, G.A. Res. 1401 (XIV), UN GAOR, 14th Sess., 842 Ple. Mtg., UN Doc. A/Res/1401 (XIV) (1959), Yearbook of the International Law Commission, vol. 2, part 2, at 49 (1974).

UN General Assembly resolution 1803 (XVII), *Permanent Sovereignty Over Natural Resources* (14 December 1962).

UN General Assembly Resolution 2669, *Progressive Development and Codification of the Rules of International Law Relating to International Watercourses*, G.A. Res. 2669 (XXV), UN GAOR, 25th Sess. UN Doc. Supp. No. 28, UN Doc. A/Res/2669 (XXV) (1970) *reprinted in* 9 ILM 1292 (1970). Resolution 2669 (XXV) was followed by resolutions 2780 (XXVI), 2926 (XXVII), 3071 (XXVIII) and 3315 (XXIX), Yearbook of the International Law Commission, vol. 2, part 1, at 150.

UN General Assembly Vote (1997), UN General Assembly resolution 51/229 dated 21 May 1997 and UN General Assembly Press Release GA/9248, available at: <http://assets.panda.org/downloads/un_watercourses_convention_voting_records.pdf> (accessed 6 July 2017).

UN Under-Secretary-General for Legal Affairs, the Legal Counsel, Letter dated 29 January 2002 (addressed to the President of the Security Council, 12 February 2002, S/2002/161) [regarding contracts signed by Moroccan authorities with foreign companies for the exploration of mineral resources in Western Sahara].

United States Agency for International Development, *Mekong, Mother of Water*, <https://www.usaid.gov/global-waters/june-2015/mekong> (accessed 23 June 2017).

United States Bureau of Reclamation, "Reconnaissance Report—Lower Mekong Basin" (1955).

United States Department of State, "Legal Aspects of the Use of Systems of International Waters," *in* Sen. Doc. No.118, 85th Cong., 2nd Sess. 43 (1958) 89–91.

University of British Columbia Institute for Resources, Environment and Sustainability, *Watershed Management*, <http://wiki.ubc.ca/Category:Watershed_Management_Certificate> (accessed 8 June 2017).

van der Leeden, Frits, *Water Resources of the World: Selected Statistics* (Water Information Center, Inc., 1975).

Vietnam National Mekong Committee (Vietnam NMC), Letter to UNDP Resident Representatives in Vietnam, UNDP, Bangkok, Thailand (12 May 1994).

Viet Nam News, *Viet Nam Vows to Preserve Water Resources, Eco-system* (7 July 2007).

Vick, Margaret J., "The Senegal River Basin: A Retrospective and Prospective Look at the Legal Regime" (2006) 46 *Nat. Resources J.* 211.

Vick, Margaret J., "International Water Law and Sovereignty: A Discussion of the ILC Draft Articles on the Law of Transboundary Aquifers" (2008) 21 *Pac. McGeorge Global Bus. and Dev. L. J.* 191.

Vick, Margaret J., *Integrated Water Law: Local to International* (unpublished JSD dissertation, McGeorge School of Law, University of the Pacific, September 2009).

Vinogradov, Serdei et al., Sergei Vinogradov et al., *Transforming Potential Conflict into Cooperation Potential: The Role of International Water Law* (UNESCO/IHP, 2003) 48, available at <http://unesdoc.unesco.org/images/0013/001332/133258e.pdf> (accessed 14 June 2017).

Vladimirov, V.B. and Paremud, S.P., "Yali Hydro Development on the Se San River in Vietnam" (1997) 31 *Hydrotechnical Construction* 39.

Wain, Barry, "Beijing's Clumsy Manoeuvres," *Far Eastern Economic Review* (26 August 2004a).

Wain, Barry, "Mekong River: River at Risk—The Mekong's Toothless Guardian," *Far Eastern Economic Review* (26 August 2004b).

Warner, Jeroen et al., "Going with the Flow: River Basins as the Natural Units for Water Management?" (2008) 10(Supp. 2) *Water Policy* 121.

Water Environment Partnership in Asia [WEPA], *Cambodia, Mekong River tributaries and rainfall, 2.1.2* <http://www.wepa-db.net/policies/state/cambodia/river2_1_2.htm> (accessed 7 June 2017).

Weatherbee, D.E., "Cooperation and Conflict in the Mekong River Basin" (1997) 20 *Studies in Conflicts and Terrorism* 167.

Weiss, Edith Brown, "Our Rights and Obligations to Future Generations for the Environment" (1990) 84 *Am. J. Intl. L.* 198.

Wells, Cameron and Piseth, Tha, *Closed Dams a Headache for Fishermen* (11 February 2010) The Phnom Penh Post <http://www.phnompenhpost.com/national/closed-dams-headache-fishermen> (accessed 18 June 2017).

White, G. et al., "Economic and Social Aspects of Lower Mekong Development" (UN Economic Commission for Asia and the Far East, Committee for Coordination of Investigations of the Lower Mekong Basin, 1962).

White, R., *The Organic Machine: The Remaking of the Columbia River* (Hill and Wang, 1995).

White, Wayne C., "A Review of the Se San 3 Hydropower Feasibility Study" (prepared for Probe International, 2000).

Willheim, Ernst, "Australia-Indonesia Sea-Bed Boundary Negotiations: Proposals for a Joint Development Zone in the 'Timor Gap'" (1989) 29 *Nat. Resources J.* 821.

Wipatayotin, Apinya, *Cambodia Eyed as New Source of Water Supply* (28 August 2007) Bangkok Post.

Wipatayotin, Apinya, *Water Grid Scheme Lashed* (1 February 2010) Bangkok Post.

Wirojanagud, Wanpaen and Inthong, Phornphimon, "Trans-boundary water quality in the Mekong River" (June 2008) *Water 21* 59.

Wisznicwski, I., "Risking Salinity in Thailand and Lao PDR" (2003) 9(1) *Watershed* 12–25.

Wohlwend, Bernard J., "The Emerging Principles and Rules of International Water Resources Law" (March 2001; paper presented at the 1st FAO Regional Training Workshop on International Water Law and Negotiation Skills in Cairo, Egypt, 25 April–1 May 2001) 9. available at <http://www.bjwconsult.com/EMERGING.PDF> (accessed 29 July 2010).

Wolf, Aaron T., *Atlas of International Freshwater Agreements* (United Nations Environment Programme (UNEP), 2002).

Wolf, Aaron T., "The Present and Future of Transboundary Water Management" in C.M. Figueres et al. (eds.), *Rethinking Water Management: Innovative Approaches to Contemporary Issues* (Earthscan, 2003a) 168.

Wolf, Aaron T., "Transboundary Waters: Sharing Benefits, Lessons Learned" (Draft Thematic Background Paper, International Conference on Freshwater, 2001) <http://lv-twk.oekosys.tu-berlin.de/project/inwent/images/pdfs/transboundary-waters.pdf> (accessed 15 June 2017).

Wolf, Aaron T., Director, Program in Water Conflict Management and Transformation, Oregon State University (email communiqué to author, 14 April 2010, 10:18 p.m. CST).

Wolf, Aaron T. et al., "International Waters: Identifying Basins at Risk" (2003b) 5(1) *Water Policy* 29.

Wolf, Aaron T., Jarvis, Tiodd, and Kunjappan, Rejani, "Effective Negotiation for Transboundary Waters: A Skills Building Course," prepared for the MRC by authors, (April 2014) 126. *Available at* <http://www.transboundarywaters.orst.edu/training/Workshop%20Materials/Mekong%20-%202014-%20Effective%20Negotiation%20TransWaters%20-%20Skills%20Building%20Course.pdf (accessed 19 September 2017).

Woodhouse, Melvin and Zeitoun, Mark, "Hydro-hegemony and International Water Law: Grappling with the Gaps of Power and Law" (2008) 10(Supp. 2) *Water Policy* 103.

World Commission on Dams (WCD), "Dams and Development: A New Framework for Decision-Making" (WCD, 2000).

World Commission on Environment and Development, "Our Common Future" (Brundtland Report, 1987).

World Rainforest Movement, *Dam in Vietnam hits Cambodians* (January 2001) Bulletin No. 42 at 1, available at <http://wrm.org.uy/oldsite/bulletin/42/Vietnam.html> (accessed 16 June 2017).

World Summit on Sustainable Development (WSSD), Johannesburg, South Africa (26 August–4 September 2002).

World Wildlife Fund, UN Freshwater Agreements Project, News Update, no. 01, at 2 (July 2007).

World Wildlife Fund, UN Freshwater Agreements Initiative, News Update, no. 07, at 3 (May 2009a).

World Wildlife Fund, UN Freshwater Agreements Initiative, News Update, no. 09, at 3 (November 2009b).

World Wildlife Fund, UN Freshwater Agreements Initiative, News Update, no. 13, at 2 (January 2011).

Worley WTL, Ltd., "Se San 3 Hydropower Project: Yali/Se San 3 Environmental and Social Impact Analysis Study" (Asian Development Bank-PPTA 31362-01-V16, RMF Draft Copy, April 2000) 2–22 to 2–23 [Worley Report].

Wormser, Michel, "Preface" in Inger Andersen et al., *The Niger River Basin: A Vision for Sustainable Management* (The World Bank, 2005).

Wouters, Patricia (ed.), *International Water Law: Selected Writings of Professor Charles B. Bourne* (Kluwer Law International, 1997).

Wyatt, Andrew B. and Baird, Ian G., "Transboundary Impact Assessment in the Sesan River Basin: The Case of the Yali Falls Dam" (2007) 23 *Int'l J. of Water Resources Development* 427.

Yevjevich, Vujica, "Water Diversions and Interbasin Transfers" (September 2001) 26 *Water International* 342.

Yearbook of the International Law Commission, vol. 2, part 1, at 302, para. 17, UN Doc. A/CN.1/SER.A/1974 (1974).

Yearbook of the International Law Commission, vol. 2, part 1, at 110 (1984).

Yearbook of the International Law Commission, vol. 2, part 2, at 25, n. 83 (1987).

Yearbook of the International Law Commission, vol.2, part 2, at 118, para. 2 (1994).

Yoffe, Shira B., *Basins at Risk: Conflict and Cooperation over International Freshwater Resources* (unpublished Ph.D. dissertation, Oregon State University, 2001), available at <http://www.transboundarywaters.orst.edu/research/basins_at_risk/index.html> (accessed 23 June 2017).

Zeitoun, Mark and Warner, Jeroen, "Hydro-hegemony—A Framework for Analysis of Trans-boundary Water Conflicts," (2006) 8(5) *Water Policy* 435.

Zhao, Michael and Schell, Orville, "Tibet: Plateau in Peril" (2000) XXV(3) *World Policy Journal* 171.

Index